ICME-13 Monographs

Series editor

Gabriele Kaiser, Faculty of Education, Didactics of Mathematics, Universität Hamburg, Hamburg, Germany

Each volume in the series presents state-of-the art research on a particular topic in mathematics education and reflects the international debate as broadly as possible, while also incorporating insights into lesser-known areas of the discussion. Each volume is based on the discussions and presentations during the ICME-13 Congress and includes the best papers from one of the ICME-13 Topical Study Groups or Survey Teams.

More information about this series at http://www.springer.com/series/15585

Murad Jurdak · Renuka Vithal
Editors

Sociopolitical Dimensions of Mathematics Education

From the Margin to Mainstream

 Springer

Editors
Murad Jurdak
Department of Education
American University of Beirut
Beirut
Lebanon

Renuka Vithal
School of Education
University of KwaZulu-Natal
Durban
South Africa

ISSN 2520-8322   ISSN 2520-8330 (electronic)
ICME-13 Monographs
ISBN 978-3-319-89189-7   ISBN 978-3-319-72610-6 (eBook)
https://doi.org/10.1007/978-3-319-72610-6

Printed on acid-free paper

This Springer imprint is published by Springer Nature
The registered company is Springer International Publishing AG
The registered company address is: Gewerbestrasse 11, 6330 Cham, Switzerland

To the memory of Christine Keitel, an internationally renowned scholar and founder of the social and political dimensions in mathematics education, for her visionary, courageous and dedicated leadership in mathematics education and for the fearless ways she used her sharp intellect to give voice and visibility to those on the margins of society, education and scholarship in mathematics education.

Preface

This monograph on *Sociopolitical Dimensions of Mathematics Education: From the Margin to Mainstream* is one of the volumes of the ICME-13 Monograph series. It includes a selection of papers presented at the International Congress on Mathematical Education (ICME-13) held in Hamburg Germany (24–31 July 2016) in Topic Study Group 34 (TSG 34) on the Social and Political Dimensions of Mathematics Education.

The main aim of this volume is to capture, document and expand the diversity of the social and political dimensions of mathematics education, issues, concerns, perspectives, contexts, and approaches presented at TSG 34 of ICME-13. There were 26 presentations in TSG 34 (4 panel papers, 18 papers in regular session, 4 oral presentations). Following the Congress, all presenters were surveyed regarding their interest to submit a paper for publication. The result showed 21 participants expressing interest in submitting chapters for consideration. All chapters that were received underwent a rigorous single-blind reviewing process. After the process of review and revision of the chapters, eventually 11 were selected to create this volume. A 12th introductory chapter was added by the editors of the volume.

The second aim of this volume is to recognize and promote the mainstreaming of the sociopolitical dimensions of mathematics education through ongoing critique and inquiry into content, policies, practices, theories, and research. This is elaborated in the introductory chapter. The third aim is to invite individual authors to (self)critique and reflect about the implications of their ideas and findings for mathematics education theories, policies, practices, activism, and research (among others) and thereby to continue moving these from the margin to the mainstream for different stakeholders and contexts.

The volume is organized in four parts in addition to an introductory chapter. The introductory chapter sets the background for the growing phenomenon of mainstreaming of the sociopolitical in mathematics education in this volume and in the field of mathematics education as reflected in ICMEs. Part I presents three chapters on theoretical perspectives on the sociopolitical in mathematics education. Part II presents two chapters on sociopolitical critiques of the role of researchers in mathematics education. Part III presents three chapters that explore sociopolitical practices in mathematics education from Brazil and Spain as well as one on a sociopolitical critique of teachers' education practices and continuous training. Part IV presents three chapters on sociopolitical critiques of media and policies in mathematics education.

We acknowledge the authors in this volume for their time and effort in undertaking the many revisions of their chapters as well as for acting as reviewers for the chapters that were submitted. We also thank Paola Valero and Belgüzar Kara for reviewing several chapters. We express our appreciation to Gabriele Kaiser, the editor of the ICME-13 Monographs Series for her support for TSG 34 and for her encouragement during the preparation of the manuscript of this volume.

This volume is dedicated to Christine Keitel, Professor Emeritus of Mathematics Education; former Vice-President of Freie University, Berlin, Germany; and former President of the Commission Internationale pour l'Étude et l'Amélioration de l'Enseignement des Mathématiques (CIEAEM). Christine Keitel passed away on 30 June 2016, just weeks before ICME-13 was held in Hamburg, Germany. Through this dedication and for this volume in particular, we acknowledge Christine Keitel as an internationally renowned scholar, founder of the social and political dimensions in mathematics education and for her contribution to several past ICMEs in this broad area of practice, research and theory. We recognize and remember her visionary, courageous and dedicated leadership in mathematics education; and the fearless ways in which she used her sharp intellect, enormous stature and distinctive academic standing in mathematics education to advance the social and political dimensions in mathematics education when doing so attracted much criticism.

It is our sincere hope that this volume successfully captures the milestone represented by the shift that has been taking place over the last four decades of moving the diverse sociopolitical dimensions from the margins to the mainstream of mathematics education.

Beirut, Lebanon   Murad Jurdak
Durban, South Africa   Renuka Vithal
September 2017

Contents

Mainstreaming of the Sociopolitical in Mathematics Education

Renuka Vithal and Murad Jurdak

Abstract This introductory chapter explores the phenomenon of mainstreaming of the sociopolitical in mathematics education. The shift of the diverse sociopolitical dimensions from the margins to the mainstream of mathematics education has been taking place over the last four decades. The chapter examines this shift by reviewing the scientific activities of the International Congress on Mathematical Education (ICMEs), conducted under the auspices of the International Commission for Mathematical Instruction (ICMI), and through the literature as it has come to express itself in this volume in comparison to the recent beginnings of the sociopolitical in mathematics education.

Keywords Mainstreaming · Social and political dimensions · Margin
Mathematics education · Rise of the sociopolitical · ICMEs

1 Introduction

While the International Commission for Mathematical Instruction (ICMI) has been in existence for over a century (since 1908) organizing and supporting all kinds of scientific activities, it is only in the last three to four decades that the sociopolitical dimensions have been recognized and developed. The programmes of the International Congress on Mathematical Education (ICMEs) hosted under the auspices of ICMI every four years, in many ways, have come to represent the state of mathematics education in a particular area and in a given period. Hence, examining the ICME programmes offers one means for observing the movement of the sociopolitical from the margin to the mainstream of mathematics education.

R. Vithal (✉)
School of Education, University of KwaZulu-Natal, Durban, South Africa
e-mail: vithalukzn@gmail.com

M. Jurdak
American University of Beirut, Beirut, Lebanon
e-mail: jurdak@aub.edu.lb

© Springer International Publishing AG 2018
M. Jurdak and R. Vithal (eds.), *Sociopolitical Dimensions of Mathematics Education*, ICME-13 Monographs, https://doi.org/10.1007/978-3-319-72610-6_1

1

The 13th International Congress on Mathematical Education (ICME-13), which took place in 2016 (Hamburg, Germany), included in one of the main activities of the congresses, the Topic Study Group 34 (TSG 34) on the Social and Political Dimensions of Mathematics Educations. Although there are other kinds of Groups such as working and discussion groups in ICME, the significance of this is that the naming of a Topic Study Groups reflects that a particular area of mathematics has been sustained and has a sufficient critical mass to generate interest from the rank and file of delegates attending any congress.

This was the first TSG on the Social and Political Dimensions of Mathematics Education in ICMEs and thereby marked an important step in the mainstreaming of this area of work as a scholarly and ongoing significant activity of the broader mathematics education community. Not only was TSG 34 a milestone, but it was also an opportunity to convene an international and diverse group of practitioners and scholars with a wide range of interests, issues, concerns and approaches in the social and political dimensions of mathematics education. The discussions during the TSG 34 sessions were intense and showed a wide variety of perspectives and that connected the sociopolitical dimensions to several other areas both inside and outside of mathematics education.

An innovation of ICME-13 was the invitation to the TSGs to publish a state-of-the-art survey of the topic before the meeting. The breadth and diversity of this area of mathematics education is discernible in the TSG 34 pre-ICME-13 publication of a topical survey on the *Social and Political Dimensions of Mathematics Education: Current Thinking* (Jurdak et al. 2016), which examines issues in five critical social and political areas in mathematics education:

- Equitable access and participation in quality mathematics education: ideology, policies, and perspectives
- Distributions of power and cultural regimes of truth
- Mathematics identity, subjectivity and embodied dis/ability
- Activism and material conditions of inequality
- Economic factors behind mathematics achievement

The above publication was an important initiative of the ICME-13 organisers and illustrates both the rise of the sociopolitical in mathematics education and its shift from the margin to the mainstream. In this introductory chapter, this main-streaming of the sociopolitical dimensions in mathematics education is identified and explored from its origins, in contrast to its current manifestations, through the scientific activities of ICMEs; and through the literature as it comes to express itself in this volume as this area of work has developed. It shows not only the movement of the sociopolitical from the margin to the mainstream in mathematics education but also its growth in diverse aspects of practice, theory, research and policy.

2 Mainstreaming of the Sociopolitical in Mathematical Education

To mainstream, according to its dictionary meaning, is to normalise, to make that which is being mainstreamed, conventional, typical or ordinary. The social and political dimensions of mathematics education or the sociopolitical in mathematics education, as this area has more recently come to be characterised, is seriously making its mark and inroads into mathematics education globally in different parts of the world, especially in this twenty first century (Gutiérrez 2013; Jurdak 2014; Straehler-Pohl et al. 2017). This can be observed in various mathematics education conferences, sole authored and edited volumes, dedicated sections in international handbooks, special issues of journals and a broad range of both scholarly publications and popular media articles as well as in thesis of postgraduate degrees and in courses for practitioners and researchers in mathematics education.

The nature and content of the sociopolitical in mathematics education is as diverse as the perspectives and aspects that are coming to constitute this area of work. It is in many ways being construed as a broad umbrella of issues and interests given by different contexts and their social, cultural, political and economic histories and as these unfold in contemporary society and a globalised world.

It is possible to discern at least two ways in which the mainstreaming or the institutionalising of the sociopolitical in mathematics education is taking place. The first can be observed in the sociopolitical dimensions being named as such and developed in their own terms as specific and dedicated areas, themes or strands in mathematics education. The second, is the way in which the sociopolitical is being infused with or integrated into many other areas of mathematics education, as existing explanations and studies fail to account adequately for what is being observed or experienced. These are not mutually exclusive with much border crossings in theory, research and practice.

This volume contributes to the mainstreaming of the sociopolitical in mathematics education by bringing voices from the periphery into the centre to influence and impact thinking and actions. However, the mainstreaming of the social and political dimensions may be achieved in a variety of directions.

One direction is to broaden the arguments for the sociopolitical in mathematics education for particular contexts of educational institutions and students described as being variously on the margin to the contexts of mainstream schools and students. The chapters in this volume demonstrate that sociopolitical awareness is evident across a range of contexts including, more affluent contexts, institutions and classrooms to create caring and ethical societies. The mainstreaming in this direction eventually leads to the question of how a critical, sociopolitical oriented mathematics education can be advanced across settings.

Another direction in which mainstreaming of the social and political dimensions may be achieved is to orient and enhance research on the sociopolitical dimensions of mathematics education that is grounded in the policies and practices of existing systems of education within particular socio-cultural contexts. Many chapters in this volume illustrate doing research in these aspects of mathematics education that is grounded in current dominant systems of education but disrupts these to reveal inequities and distribution of power.

In this volume, the chapters show how diverse research questions, methods and theories, relevant to sociopolitical dimensions, are being critically appropriated and advanced by growing numbers of mathematics educators because of their value and appropriateness for explaining broader social, economic, cultural and political realities in which mathematics education is enacted.

These forms of mainstreaming of the sociopolitical in mathematics education can be demonstrated by contrasting programmes of ICMEs, for instance, the first time the sociopolitical dimensions of mathematics education were acknowledged in a special fifth day programme in ICME-6 (1988) compared to its more pervasive manifestation in the programme of ICME-13 (2016). This is explored in the next section.

3   Mainstreaming of the Sociopolitical in ICMEs' Activities

The International Congress on Mathematical Education (ICMEs), which take place every four years, is a meeting that most scholars in mathematics education aspire to attend. In many ways, ICME programmes represent the current dominant thinking, concerns and issues in mathematics education, of a particular moment and period. From this perspective, organizationally and by its activities, ICMEs can themselves be analysed sociopolitically, especially since thousands of delegates from a vast majority of countries from around the world attend each ICME. What features in the mainstream programme and what gets onto the periphery are decisions made by the respective International Programme Committee and indirectly, the leadership of the ICMI Executive. However, equally, delegates shape what is considered important and urgent, and what is not, by their attendance and participation.

It was quite literally, at the margins of an ICME that attention was first drawn globally, to the sociopolitical dimensions of mathematics educations. A special fifth day was added to the programme of ICME-6 in 1988 (Budapest, Hungary) to address a range of social, political and related issues and concerns. That the moment had arrived for these ideas was reflected in the large number of 90 contributors from over 40 countries to this programme. Even though it was placed at the periphery of the Congress towards the end, it nevertheless also took centre stage as a whole day event devoted to diverse sociopolitical and related aspects from all over the world. Keitel (1989), who was the chief editor for the proceedings that followed as a UNESCO publication with the title *Mathematics, Education and Society*, explained the intention and rationale in the introduction to the section on "Policy Dimensions of Mathematics Education Research and Practice" in the *Second International Handbook of Mathematics Education*

> ... (it) indicated an increasing awareness of education in general as a universal human right and mathematics education in particular...(and) represented an intent to investigate the interrelationship between mathematics education, educational policies and social and cultural conditions in a broad sense. It was accepted for the first time as a legitimate challenge, a matter of worldwide consciousness and recognition....

> The outcomes of this 5th day special program constituted more of an agenda for future activities than a balance account of achievements and limitations of mathematics education under present social and political conditions. However, the message could be disseminated and partly implemented as a necessary complement to future activities within ICME and other conferences of mathematics educators. (Keitel 2003, p. 3)

This first ICME programme that focussed primarily on the social, cultural and political dimensions of mathematics education, was organized around four broad themes, which give an indication of the range of issues being engaged:

- Mathematics education and culture
- Society and instititionalised mathematics education
- Educational institutions and the individual learner
- Mathematics education in the global village

In successive ICMEs that have followed, different aspects of the social and political dimensions have been taken up and dealt with both explicitly and implicitly in plenary presentations and panels, invited and regular lectures, topic study groups, working and discussion groups, thematic and national presentations; survey teams and ICMI studies, and through ICMI affiliated organisations. These include a diversity of areas such as gender, class, culture, language, identity and all kinds of (in)equities, to name but a few, which have been focused on over the past three to four decades. This fifth day special programme represents the first form of mainstreaming set out above and has continued inside but also outside ICME, for example, in the international Mathematics Education and Society (MES) conferences (for references see Jurdak et al. 2016).

In contrast to ICME-6 and that first fifth-day special programme in 1988, an analysis of the ICME-13 programme in assessing how far concerns with the social and political dimensions have moved into the mainstream of the programme, reveals quite a different picture in the year 2016, in the historical trajectory of ICMEs. A pervasive infusion of a concern with a wide variety of sociopolitical dimensions and related to cultural, economic, language and broader context matters, may be observed in the ICME-13 programme. It is possible to analyse engagement with social and political aspects in the *ICME-13 Final Programme,* as a data source, across almost all ICME-13 programme activities.

In most of the plenary lectures reference was made to some or other sociopolitical aspect. At least four of the six plenary lecturers explicitly refer to the sociopolitical in some form in their abstracts. For instance, Barton's talk on *Mathematics, education and culture: a contemporary moral imperative* offered reflections on current societal cross-cultural crises. The sociopolitical aspects were clearly discernible by invoking the first ICME plenary that brought culture into the mainstream, namely, D'Ambrosio's plenary lecture in ICME-5 (1984) on the *Socio-cultural bases for mathematics education* and subsequent developments in ethno-mathematics, which are referred to as the "historical, political and cultural dimensions and their relation to society" (p. 17). Similarly, Ball's plenary lecture included a sociopolitical discourse in seeking to "support classrooms as equitable communities of practices" (p. 18); and Ziegler included the question of the relevance of "the public image of the subject... and thus ultimately determines what mathematics can achieve, as a science, as a part of human culture, but also as a substantial component of economy and technology" (p. 17). The plenary panel on *International comparative studies in mathematics* identified as one of its lessons a focus "on policy in local contexts" (p. 16); and a second panel on "transition in mathematics education" examined "transitions between social groups or contexts with different mathematical practices" (p. 19).

Furthermore, the ICME-13 programme shows that three of the five invited lectures of the prestigious ICMI Awardees referred to and were recognized for their work related to the broader social, cultural and political dimensions of mathematics education. While Leung asked "Does culture matter?" in making sense of East Asian mathematics achievement in his lecture, Bishop's topic continued his three-decade long engagement with cultural perspectives and Adler was recognized for her multilingualism work (p. 15).

In addition to the Topic Study Group specifically named "Social and Political Dimensions of Mathematics Education" (TSG 34) for the first time, at least a quarter of the 54 Topic Study Groups on the programme could be regarded as being part of TSG 34 or related to it. These include topics such as equity (including gender); multilingual and multicultural environments; affect, belief and identity; mathematics education in and for work; popularization of mathematics; interdisciplinary mathematics education; language and communication; mathematical literacy; the role of history of mathematics in mathematics education; the role of ethnomathematics in mathematics education; and a TSG on diversity of theories in mathematics education.

Another set of activities in the ICME-13 Final Programme which show an engagement with sociopolitical dimensions was in the ICMI Studies and Survey Teams. The study team examining distance, blended and e-learning in mathematics, for example, included a question of: "what are the cultural, economic and political questions to be aware of as different countries experience different degrees of internet driven changes in mathematics education" (p. 26). The study on teachers working and learning through collaboration, refer in their abstract, to concerns about teachers not only knowing mathematics "but also able to demonstrate... social and ethical knowledge in working with their students at any level. They have to work according to societal, political and institutional demands which shape and challenge their professional, personal, social and cultural identities" (p. 27).

A further political reading of the ICME-13 programme that can be made is in the diversity observed in the selection of chairs, presenters and a range of leadership roles assigned to be inclusive of all parts of the globe and across different divides of culture, language and nationalities. The category of National Presentations in the ICME-13 programme reflected a broad range of countries or regions, which opens opportunities for social and political aspects to be brought to the fore. National presentations were made at ICME-13 by Argentina, Brazil, Ireland, Japan, Turkey and the Lower Mekong Sub-Region (pp. 34–5).

International bodies such as ICMI and the organising of large scale international programmes and events like international congresses are in themselves social and political; and therefore, are under constant critique and scrutiny for how they change in order to respond to different challenges and impetuses in international and global environments and among those with interests in mathematics education. It was, perhaps, not surprising that following some criticisms of the special fifth-day programme of ICME-6 (described above) and a recognition of the need for a much more dedicated focus on the political dimensions of mathematics education, that the first international conference on the Political Dimensions of Mathematics Education (PDME) was hosted soon after in London in 1990. Two further PDME conferences were hosted before being renamed Mathematics Education and Society (MES) conferences, and a further 9 such conferences have been hosted and proceedings produced (see Jurdak et al. 2016 for all PDME and MES proceedings references).

It can be observed, however, that the MES conferences have never been recognized or affiliated with ICMI as an organization, compared to, say, the International Organisation of Women and Mathematics Education (IOWME), which is an affiliated organization. The politics of establishing IOWME is a different earlier historical episode involving strong activism by women delegates. At ICME-7 in 1992 (Quebec, Canada), after the ground breaking ICME-6 special programme that launched the sociopolitical dimensions of mathematics education, an attempt was made to expand one of the strong and large affiliated organisations, the International Group for the Psychology of Mathematics Education (PME), by arguing for changing the "P" in PME to "R" for Research to be more accommodating and inclusive of social and political dimensions. Although this initiative was unsuccessful, subsequent conferences of PME have endeavoured to integrate an engagement with sociopolitical dimensions in their conferences and proceedings. Moreover, this also spurred on and led to the establishment of the Mathematics Education and Society conferences. In this way, the two forms of mainstreaming may be seen to be connected.

It is evident that the recognition and content of the social and political dimensions of mathematics education have continued to grow and develop organisationally, both within ICMI and ICMEs; and outside, in many other spaces inhabited by mathematics educators such as the MES conferences. The ICME-13 programme shows how the sociopolitical dimensions have moved from the margin into the mainstream by: firstly, being grown in their own right as an important area of practice and scholarship; and secondly, becoming integrated into a range of different activities and into the different areas of mathematics education as they are increasingly being engaged.

4   Mainstreaming of the Sociopolitical in Mathematics Education in the Literature

The words "politics" and "mathematics education" first appeared together in the title of the seminal work of Mellin-Olsen in *The Politics of Mathematics Education*, which was first published in 1987 and signalled that the sociopolitical dimensions had begun to make their mark in the mainstream literature of mathematics education (at least in the English language).

> ...For many the world in 1985 was not a peaceful place.
>
> In mathematics education we have arrived at a stage where we discuss personal knowledge, shared knowledge, the need for pupils to develop their own mathematical ideas and tools in order to gain some insight into the power of mathematics.
>
> It is difficult to see how personalized knowledge can be discussed without using notions such as politicization, conflict and oppression when the potential learner may be in the midst of bitter struggles for civil rights. At the same time it is not exaggerated to say mathematics educators have not been at the forefront when it comes to politicizing education.
>
> ...I attempt to build a general theory of the politicization of mathematics education... this book is the result of a twenty-year long search to find out why so many intelligent pupils do not learn mathematics, whereas at the same time, it is easy to discover mathematics in their out-of-school activity. (Mellin-Olsen 2002, p. xiii)

Although the earliest writings in this area can be traced to the 1970s in different parts of the world with publications not only in English (for references see Greer and Skovsmose 2012), the decade of the 1980s brought the political dimensions of mathematics education to the surface in sharp relief internationally; and coincided with other developments such as in social, cultural, gender, and equity aspects, to name but a few. Keitel captured this diversity in sociopolitical issues when describing the papers presented as part of the special fifth day programme of ICME-6 (1988) that sought to address the varied contexts of mathematics education:

> One important focus was on analyzing conditions and causes of the restricted teaching and learning opportunities for pupils of certain groups defined by gender class and ethnical minority in industrialised countries as well as for the majority of young people growing up in the non-industrialised "Third World". The community of mathematics educators agreed to search for the means to overcome Eurocentrism and cultural oppression in mathematics teaching and learning, and in the design of curricula, learning materials, learning environments and to adopt critical and multicultural perspectives which allow meaningful mathematics learning to be related to social experiences and social needs. (Keitel 2003, p. 3)

These foundational reflections on the sociopolitical dimensions highlight from the outset the early preoccupations, although interested and concerned with action, were seized with the search for theories and explanations. It is not surprising then that the broad range of concerns and issues that are variously linked to the sociopolitical dimensions, have drawn equally, on a wide diversity of theories and perspectives from both inside and outside mathematics education.

The chapters in this volume on *Sociopolitical Dimensions of Mathematics Education: From the Margin to Mainstream* demonstrate this breadth of theoretical works being engaged, such as Foucault, Bourdieu, Lacan, Deleuze, Engeström, Luhmann, Bernstein, Freire, Skovsmose, Apple, Butler, Edelman and many others. Theoretical developments, however, as a primary concern in the sociopolitical area, have experienced a shift toward situating the sociopolitical in mathematics education within broader theoretical frameworks by drawing on and integrating diverse theoretical perspectives. The chapter by Jurdak attempts to integrate the sociopolitical and sociocultural dimensions by seeking connections between notions of power and culture. Lensing and Straehler-Pohl, however, focus on a particular theoretical question in their chapter, that of conceptualizing an ethical perspective for mathematical applications in contemporary society. Appelbaum, in his chapter, offers three sets of nomadic epistemological categories and argues that such categories can change our worlds of possibility for mathematics education while allowing coexistence with more mainstream programs of research and practice. Moreover, this conception enables mathematics education to be viewed as an alterglobal social movement.

While theoretical and conceptual tools are important developments in and of themselves and are needed for action, they have to be interpreted for a range of practices—in mathematics classrooms, for mathematics curricula and in all their respective components, for policy and for research.

In mathematics education research, the sociopolitical dimensions have opened for serious and new questions about research methods, methodologies and in respect of the roles and responsibilities of researchers themselves as well as for their participants (e.g. Valero and Zevenbergen 2004; Straehler-Pohl et al. 2017). This volume adds to significant literature that has emerged in this area. Two chapters in this volume demonstrate an engagement with the role of the researcher in undertaking studies in the sociopolitical dimensions of mathematics education. The chapter by Darragh examines researcher positioning and the power inherent in research activity through the recognition of identity work on the part of the researcher and in how participants are implicated in the (re)production of particular social and political discourses. Keeping the gaze on researchers, in his chapter Pais implores those who study mathematics education to reflect on themselves as part of the problem in how common ideological assumptions or truths about mathematics educations are naturalized and propagated through mechanisms of power, and which make it difficult for researchers to see beyond themselves.

Since the emergence of sociopolitical discourses, there have been a wide variety of practices described and studied to shed light on how the intentions and goals associated with sociopolitical perspectives could be realized for students and teachers in different contexts (e.g. Skovsmose 1994; Vithal 2003; Gutstein 2006). Three chapters in this volume exemplify the broadening of sociopolitical mathematics education as a field of research and application. The chapter by Campos, Hess and Sena in this volume explores a critical financial education, showing the intense student interest and involvement in an undergraduate financial mathematics course when elements of a critical mathematics education were integrated into a modelling activity. The focus shifts from students to teachers in the chapter by Bruno, Ruiz-López and de Castro. In their case study of a mathematics teacher who explicitly identifies himself as an educator for social justice, they shed light on understandings of the disjuncture between the teacher's declared ideals and his practice. Mathematics education practitioners—be they teachers, teacher educators, researchers, policy makers or analysts—are implicated in these (re)production of discourses and actions that they participate in, especially when examined through sociopolitical framings. The argument made by Montecino about the permanently outdated mathematics teacher as an outcome of the market, shows how researchers and teacher educators are implicated in the continuous elaborations of new techniques, practices and knowledge that the mathematics teacher is expected to acquire to be considered successful and to stay in the system.

Arguably, the least developed area in mathematics education literature, notwithstanding the growth of a substantial literature in sociopolitical dimensions, is the broad area of what may be referred to as mathematics education policy studies and that which examines relations among policy, practice, research and theory (Lerman et al. 2002; Vithal and Volmink 2005). It could be further argued that no matter the power of theories or successful small case studies in advancing understandings and explanations about different aspects of teaching and learning mathematics, it is national and/or institutional educational and related policies and their implementation, (as these are interpreted), which directly impacts and shapes the majority of lives and experiences of students and teachers on the ground. However, policies and practices, implicitly or explicitly, embed particular theories and perspectives. Three chapters in this volume indicate the increased engagement of the sociopolitical in mathematics education with issues of policies in actual broader contexts. In her chapter, Meaney argues that while discussions about access, inequality, power and identity have become prominent in mathematics education curricula, they are simultaneously being subordinated to neoliberal discourses of competition and accountability. Adams and Povey show how teacher autonomy has changed historically through a study that juxtaposes teacher stories from a curriculum development project in the 1970s and 1980s with those from currently serving teachers. They identify possibilities for resistance to neoliberal political agendas, which they argue, dominate educational policies and practices. The chapter by Lange and Meaney, in which they analyse and critique the "common sense" arguments presented by a Minister for Education in the media, reveals not only a shift in approach to early childhood education in Norway but also raises questions about the use of media to make educational changes in contemporary society. The role of media, especially social media opens a completely new, yet to be explored area, of how public understandings and public demands for and in mathematics education are being shaped and which in turn, influence politicians and policy makers, who are much more sensitive and responsive to these than arguably, to mathematics education experts or scholars.

5 Concluding Remarks

The rise of the social and political dimensions and its shift from the margin to the mainstream over several decades has been characterized, earlier, as a "social turn" (Lerman 2000) in mathematics education, and more lately, in emphasizing the political dimensions, as a "sociopolitical turn" (Gutiérrez 2013). This growth in theorizing, research and practices exploring the sociopolitical dimensions in mathematics education has also been accompanied with ongoing critique, has spawned a wide variety of writings and brought diverse disciplinary interests to bear on the teaching and learning mathematics broadly as attested to in this volume.

No doubt challenges to and within the sociopolitical dimensions will continue to ferment and be generative of ideas, as a recent provocatively titled publication "*Disorder of Mathematics Education*" exemplifies. Straehler-Pohl et al. (2017, p. 1) "identify and conceptualise "disorder" as the foundation of the sociopolitical dimensions and accordingly propose a shift from focusing on diversity toward focusing on disorder". They challenge mathematics educators, for example, to consider that their very search to achieve one of the core ideals of the sociopolitical in mathematics education, that of "mathematics for all", is an illusion that allows the status quo of its systematic failure to remain. These editors (some of whom have chapters in this volume), in recognising that the sociopolitical dimensions have become instituionalised as one of the important established strands in mathematics education, critique precisely this process for the tensions and contradictions it produces.

It could be argued, given the growth of the sociopolitical in mathematics education and its take-up in other areas that so far, only the tip of an iceberg has been explored. There are still many gaps and silences that need addressing as the sociopolitical dimensions advance in mathematics education (see for e.g. Jurdak et al. 2016). Some big "P" questions of relations in policy, politics and power remain: what theories, practices and research speak to those in power who make decisions, and which impact the vast majority of teachers and students? Put differently, do those with political power and access to levers on state resources to effect real changes in mathematics education at macro levels, pay attention to the careful scholarship and rigorous publications in mathematics education? Or are mathematics educators preoccupied with speaking mainly to themselves as a community? Mathematics educators participate and are implicated as practitioners in their many different roles and responsibilities: from being teachers in elementary school to those who participate and have responsibility in state institutions for developing mathematics curricula and their implementation. While a great deal is known about the former, much less is known about the latter.

In returning to the point made by Mellin-Olsen more than three decades ago (see above: "it is not exaggerated to say mathematics educators have not been at the forefront when it comes to politicizing education"), begs a different but related question: Are mathematics educators with expert knowledge and understanding at the forefront shaping and influencing decisions about mathematics teaching and learning in their respective countries or regions? If not, why not? Much remains to be done in taking the scholarship and insights from the sociopolitical dimensions of mathematics education into those spaces of power and politics that have real effects societally and to bring those spaces of power and politics under a research gaze in the sociopolitical explorations of mathematics education.

References

Greer, B., & Skovsmose, O. (2012). Introduction: Seeing the cage? The emergence of critical mathematics education. In O. Skovsmose & B. Greer (Eds.), *Opening the cage: Critique and politics of mathematics education* (pp. 1–20). Netherlands: Sense Publishers.

Gutiérrez, R. (2013). The sociopolitical turn in mathematics education. *Journal for Research in Mathematics Education, 44*(1), 37–68.

Gutstein, E. (2006). *Reading and writing the world with mathematics: Toward pedagogy for social justice.* New York: Routledge.

Jurdak, M. (2014). Socio-economic and cultural mediators of mathematics achievement and between school equity in mathematics education at the global level. *ZDM—The International Journal on Mathematics Education, 46,* 1025–1037.

Jurdak, M., Vithal, R., de Freitas, E., Gates, P., & Kollosche, D. (2016). *Social and political dimensions of mathematics education: Current thinking.* Switzerland: Springer Open.

Keitel, C. (Ed.). (1989). *Mathematics, education and society.* Science and Technology education. Document Series No. 35, Division of Science Technical and Environmental Education. Paris: UNESCO.

Keitel, C. (2003). Section 1: Policy dimensions of mathematics education research and practice introduction. In A. J. Bishop, M. A. Clements, C. Keitel, J. Kilpatrick, & F. K. S. Leung, (Eds.), *Second international handbook of mathematics education* (pp. 1–7). Dordrecht: Kluwer Academic Publishers.

Lerman, S. (2000). The social turn in mathematics education research. In J. Boaler (Ed.), *Multiple perspectives on mathematics teaching and learning* (pp. 19–44). Westport, CT: Ablex.

Lerman, S., Xu, G., & Tsatsaroni, A. (2002). Developing theories of mathematics education research: The ESM story. *Educational Studies in Mathematics, 51*(1–2), 23–40.

Mellin-Olsen, S. (2002). *The politics of mathematics education.* Dordrecht: D Reidel Publishing Company.

Skovsmose, O. (1994). *Toward a critical philosophy of mathematics education.* Dordrecht: Kluwer Academic Publishers.

Straehler-Pohl, H., Bohlmann, N., & Pais, A. (Eds.). (2017). *The disorder of mathematics education: Challenging the sociopolitical dimensions of research.* Switzerland: Springer.

Valero, P., & Zevenbergen, R. (Eds.). (2004). *Researching the sociopolitical dimensions of mathematics education: Issues of power in theory and methodology.* Dordrecht: Kluwer Academic Publishers.

Vithal, R. (2003). *In search of a pedagogy of conflict and dialogue for mathematics education.* Dordrecht: Kluwer Academic Publishers.

Vithal, R., & Volmink, J. (2005). Mathematics curriculum roots, reforms, reconciliation and relevance. In R. Vithal, J. Adler, & C. Keitel (Eds.), *Researching mathematics education in South Africa: Challenges and possibilities* (pp. 3–27). Pretoria: Human Sciences Research Council.

Part I
Theoretical Perspectives on the
Sociopolitical in Mathematics Education

Integrating the Sociocultural and the Sociopolitical in Mathematics Education

Murad Jurdak

Abstract The purpose of this chapter is to seek an integration of the sociocultural and sociopolitical perspectives in mathematics education by integrating a locally attuned version of Bourdieu's field theory (Ferrare & Apple in Camb J Educ 45 (1):43-59, 2015) and activity system (Engeström in Learning by expanding: an activity-theoretical approach to developmental research. New York: Cambridge University Press, 2015) to disrupt the separate development of the two perspectives. I combine the two theories using modular integration. Next, the chapter discusses the implications of this integration to mathematics education research, practice, and policies. I conclude with a personal narrative on my theoretical journey to sociopolitical mathematics education.

Keywords Activity theory · Bourdieu field theory · Mathematics education
Sociocultural · Sociopolitical · Integration

1 Introduction

Up to the 1960s, the social dimension of the mathematics education discourse had witnessed a recurring tension between school mathematics for social utility and school mathematics for the intellectual development of the individual. The sixties of the past century represented the climax of a movement that considered school mathematics as a cornerstone for not only the intellectual development of the students but as a necessary tool for student academic progression and as a basis for science and technology and hence for the socioeconomic development of countries.

One example, that of Shulman (1970), reflects the dominant conception of teaching mathematics in the sixties of the last century. Shulman's model (Fig. 1) featured teaching as a one-directional system in which the input (entering

M. Jurdak (✉)
American University of Beirut, Beirut, Lebanon
e-mail: jurdak@aub.edu.lb

© Springer International Publishing AG 2018   15
M. Jurdak and R. Vithal (eds.), *Sociopolitical Dimensions of Mathematics Education*, ICME-13 Monographs, https://doi.org/10.1007/978-3-319-72610-6_2

characteristics of learners) is to produce output (objectives of instruction) through the process of instruction. The emphasis on subject matter and teaching are obvious from the examples given in Fig. 1. Also, the interaction among "type of subject matter," "type of instruction," and "amount and sequence of instruction" is not made explicit in this system. Notably, no mention is made of the role of the broader socioeconomic and cultural context of teaching and learning.

By the 1980s mathematics education witnessed what Lerman (2000) termed a *social turn* in mathematics education to refer to a paradigm shift in the conception of teaching and learning by stipulating that learning and teaching are products of social activity. The first phase of the social turn recognized the social and school material contexts as core components of instruction. It considered the context as a "given," which may constrain or support the components of the system contextual dimension to teaching. However, the social turn did not challenge the one-directional system which assumes that the process of instruction acts on the input (entering characteristics of learners) to produce the output (instructional objectives). According to Cobb (2006), the social turn was characterized by approaches, which "accounted for learning in terms of internal cognitive processes, but acknowledge that cognition is influenced by social interactions with others and, to a lesser extent, by the tools that people use to accomplish goals" (p. 189). Vygotsky's assertion (1978) that cultural products, like language and other symbolic systems, mediate thought marked a shift in psychology from treating cognition and culture as separable to the view that "cognition and culture are no longer regarded as divisible" (Lucariello 1995, p. 1).

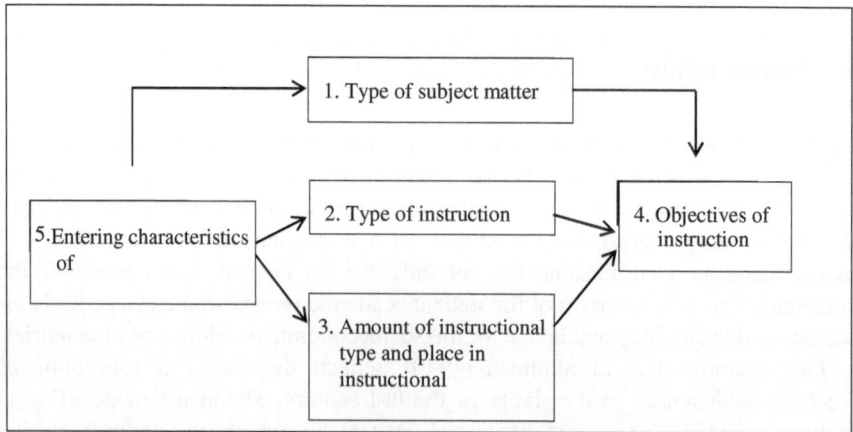

1   Mathematics, foreign languages, social studies (subject matter is defined in task terms)

2   Expository-discovery (degree of guidance); inductive-deductive

3   Number of minutes or hours of instruction; position in sequence of instructional types

4   Products; processes; attitudes; self-perceptions

5   Prior knowledge; aptitude; cognitive style; values

Fig. 1 Theoretical generalization about the nature of instruction (Shulman 1970, p. 63)

Later, Lerman (2006) introduced the term *strong social turn* to refer to a theoretical trend that argued for "the situatedness of knowledge, of schooling as social production and reproduction, and of the development of identity (or identities) as always implicated in learning" (p. 172). The strong sociocultural approaches followed Vygotsky's argument that social and cultural processes do not merely condition internal cognitive processes, but rather form learners' minds as they engage in social and cultural practices. Skovsmose and Greer (2012) describe the social turn as "manifesting the humanization (a word frequently used by Freire) of mathematics and mathematics education, encapsulated in the phrase "mathematics as a human activity" (p. 4).

In contrast to the sociocultural developments, Gutiérrez (2013) used the term *sociopolitical turn* to signal "the shift in theoretical perspectives that see knowledge, power, and identity as interwoven and arising from (and constituted within) social discourses" (p. 40). The political turn in education was ushered by the Marxist idea that the production and transmission of knowledge serve the interests of the ruling class that controls the means of material production. Until the eighties of the past century, the prevailing idea was that mathematics learning is universal and does not lend itself to the Marxist idea. The political dimension of mathematics education grew out of the realization that mathematics, as enacted in schools or produced by research, is not immune from the power structure and its distribution in the society. That idea led to a growing awareness that economically and socially advantaged groups created and maintained a schooling system that systematically favors students with privileged backgrounds by adopting a seemingly class-neutral education. The perception of mathematics as a formal language capable of imparting any meaning, promoted the idea that mathematics may act as an instrument for exercising power by imposing meanings on students. Two of the leading sociopolitical theories in general education included emancipatory education (Freire 1970/2013) and Bourdieu's field theory (Bourdieu and Passeron 1990). Skovsmose's (2011) critical mathematics education was one of the leading sociopolitical in mathematics education. The political in mathematics education was described by Skovsmose and Greer (2012) as (re)humanizing mathematics and mathematics education, which "are inextricably political activities" (p. 4).

The literature indicates that the sociocultural and the sociopolitical perspectives in mathematics education have been generally developing along separate paths. Following a mathematics education review, Pais (2012) concludes that exclusion and inequity "within mathematics education, and education in general, are integrative parts of schooling and cannot be conceptualised without understanding the relation between scholarised education and capitalism as the dominant mode of social formation" (p. 51). By pushing this argument to its extreme, he arrives at a deadlock which states "that exclusion is something inherent to the school system we realise that to end exclusion means to end schooling as we know it" (p. 82). My premise is that there are theoretical tools to disrupt the separate development of the sociocultural and the sociopolitical perspectives in mathematics education.

The purpose of this chapter is to disrupt the separate development of the sociocultural and the sociopolitical perspectives in mathematics education by integrating a locally attuned version of Bourdieu's field theory (Ferrare and Apple 2015) and Engeström's (2015) activity system. I use a modular integration approach (Markovsky et al. 2008) to bring together the two theories. Modular integration "treats two or more theories as integral modular components that can be used separately or jointly, as needed, much like different modularized electronic components can be used either alone or together in an integrated circuit for specific applications" (Kalkhoff et al. 2010, p. 3). This integration brings the two theories together to construct a common theoretical foundation for sociopolitical mathematics education. Each theory by itself does not explicitly explain the complexity and interaction of social and political dimensions of mathematics education. On the one hand, Bourdieu's field theory, which views the process of education from the perspective of power through cultural reproduction, does not address education as a socialization/acculturation developmental process. On the other hand, CHAT, which views education as a developmental collective purposeful activity embedded in a sociocultural context, is silent on the issue of power in the educational field.

The rationale for seeking an integration of the separate development of the sociocultural and the sociopolitical perspectives in mathematics education by integrating a locally attuned version of Bourdieu's field theory (Ferrare and Apple 2015) and Engeström's (2015) activity system is to disrupt the separate development of the two perspectives. On one hand, sociocultural theories view school mathematics education as a collective human activity whose purpose is to engage students in socially and culturally relevant mathematical experiences. On the other hand, sociopolitical theories propose that in school mathematics education, advantaged social groups have the opportunity to exchange and disguise their possession of different forms of power to dictate policies, practices, and beliefs in mathematics education institutions and actors, thus reproducing inequities and exclusions in mathematics education. The chapter proposes a theoretical way to disrupt the separate development of the sociocultural and the sociopolitical in mathematics education by integrating the two theories of a locally attuned version of Bourdieu's field theory (Ferrare and Apple 2015) and Engeström's (2015) activity system. The integration of the two theories is intended to provide support to the idea that mathematics education is valued as a socially and culturally enterprise and simultaneously may help mitigate the impact of power in mathematics education by providing "a space to interrupt the arbitrary and inequitable valuation of certain cultural forms over others" (Ferrare and Apple 2015, p. 54).

In what it follows, I frame this chapter within my own epistemological interests and preferences. The focus of this chapter is, by design, mathematics education whose object is the learning and teaching mathematics, mainly in school context. My choice of activity theory as one candidate for a possible integrated sociopolitical mathematics education theory is based on my epistemological orientation as reflected in my research work. My interest in Bourdieu's field theory, which is rather recent, is motivated by a desire to intensify my engagement with the political dimension of mathematics education.

2 Engeström's Cultural Historical Activity Theory (CHAT) and Mathematics Education

According to Engeström (2001), CHAT developed in three stages: basic individual human activity (Vygotsky and Leont'ev), collective human activity (Leont'ev and Engeström), and interacting activity system (Engeström). The first generation was ushered by Vygotsky's (1978) idea of cultural mediation of actions, in which the conditioned direct connection between stimulus (S) and response (R) is transcended by "a complex mediated act". Vygotsky's idea of cultural mediation of actions is commonly represented as a triangle with subject, object, and mediating artifact as vertices (Fig. 2)

2.1 Individual Human Activity

According to Leont'ev (1978, 1981), the individual human activity involves a person or group of persons who engage in an action, using tools (artifacts), to achieve an outcome embodying the intended goal. Leont'ev distinguishes three activity-related concepts: activity, actions, and operations—and he relates these concepts respectively to the motives, goals, and conditions under which the activity occurs.

The starting point in any activity is that the person who engages in the activity should have a motive, without which the activity fails to initiate. According to Leont'ev (1978), "unmotivated" activity is an activity in which the motive is not subjectively and objectively explicit. The motive is concretely translated into a possible achievable goal. The desire to achieve this goal generates actions, which are not random but subordinated to a conscious purpose on the part of the person engaged in the activity. Just as the concept of activity is subordinate to the concept of motive, the concept of action is subordinate to the concept of goal. The artifacts (material and symbolic tools) that are accessible under the objective conditions of the specific social-cultural context mediate the actions.

Fig. 2 Individual human activity

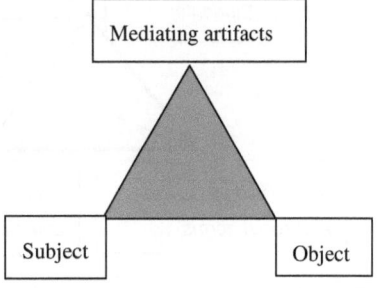

The activity of learning mathematics in school is an exemplar of individual activity. The learner, motivated to learn mathematical competencies and concepts, takes actions to achieve the intended mathematical goal, using operations, mediated and constrained by the accessible material and symbolic artifacts that exist in the objective conditions of the social-cultural context of the school.

A core premise of activity theory is the centrality of the learner's agency in the learning activity. For the learning activity to start, the activity goal has to be meaningful to the learner to evoke engagement in taking action toward realizing the goal. The choice of actions as well as the artifacts is contingent on the learner's choices and consciousness of the potential effectiveness of the actions in realizing the intended goal.

2.2 Collective Human Activity

The second generation was ushered by Leont'ev (1981) who expanded the concept of individual activity to a collective activity by introducing the element of division of labor as an essential component of collective activity. Engeström (2015) formally introduced and represented the collective activity as an activity system (Fig. 3).

The activity system is a collective purposeful activity in which a subject (or subjects) is engaged to attain an object shared by a community, using mediating artifacts, where responsibilities are assigned collectively among members of the community (division of labor) according to organizational rules and socio-cultural norms (rules). School mathematics is an example of a collective learning activity. Figure 3 represents the activity system of school mathematics.

The activity system of learning mathematics in school is a social space where students are motivated to engage in learning mathematics using appropriate

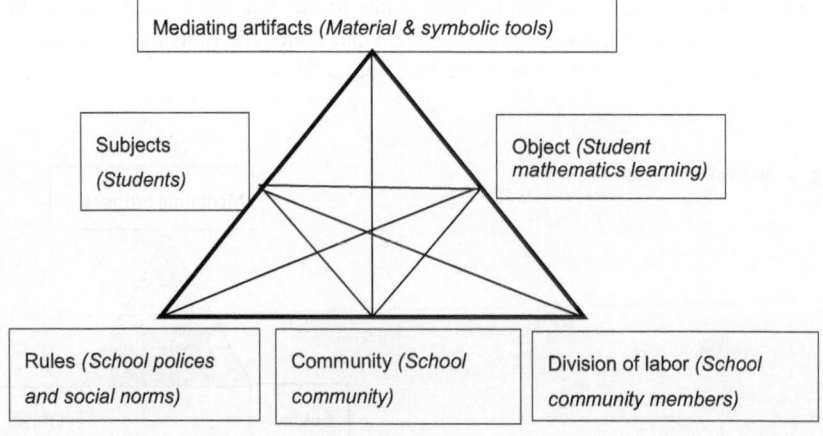

Fig. 3 School mathematics as an activity system

artifacts, which include symbolic tools such as language and mathematics as well as material tools manipulative learning materials and computers. The collective activity introduces the concept of "community" to the individual activity and thus triggers profound consequences to the dynamics of the individual learning activity. The learner in the collective activity does not assume only the identity of an individual but also that of a member of a community, which shares the same object (learning mathematics). Thus, the learner's actions and interactions are not simply individual behaviors linked to individual motives, but also moderated by the motives, actions, and interactions of the community members.

The mediating artifacts available to the learner are also communal cultural objects closely linked to the historical-cultural development of the school system. The existence of the community calls for the need for "division of labor" among the community members (student, teachers and staff, parents), and for 'rules' to govern the actions and interactions within the collective activity; regulatory rules that are explicit and public, while the social and cultural norms of the broader community of the school that are implicit and invisible.

2.3   The Role of Culture in the Activity System

Culture plays a central role in shaping the social space of school mathematics education. The school community in the activity system of school mathematics is a microcosm of the broader community, which the school serves. The social and cultural carryover from the broader community to the school community seem to affect all the nodes of the activity system of school mathematics. The process of acculturation, i.e., acquiring and appropriating the culture of the community, affects the pedagogic mode of acculturation of school mathematics. According to Jurdak (2016a), an acculturation mode of *transmission* views the pedagogic mode of teaching mainly as deficit filling. The *participation* mode of acculturation, however, views the pedagogic mode of teaching as constructing and negotiating learners' meanings. The *inculcation* mode views the pedagogic mode of teaching mainly as 'imposing' knowledge and values on students. The *mediating artifacts,* particularly the symbolic tools such as language and mathematical practices, used in learning and teaching mathematics are also cultural tools that belong to cultural contexts of the broader community. The *rules* of the activity system of school mathematics, which includes cultural and social customs and traditions of the broader school community, tend to mediate the actions of students in an invisible way. The *division of labor* in the activity system of school mathematics involves the assignment of roles of students, teachers and administrator in the teaching learning process. The division of labor in school mathematics reflects also the communal beliefs regarding authority and individual agency. Such beliefs range from a total respect to authority to a recognition of free individual agency. The object of the activity system is not only an individual goal but also a communal object, determined by the broader community in the form of a curriculum or standards.

2.4 Mathematics Education Research and CHAT

During the last two decades, the use of activity theory in mathematics education research has intensified. Almost all studies focused on the impact of sociocultural contexts on mathematics learning with little engagement in the role of power in such contexts. Stone and Gutiérrez (2007) conducted a study involving an instructional design in which multi-aged participants who are members of various cultural communities are encouraged to coordinate their efforts on educational tasks. The study examined how the multiple activity systems of the undergraduate course and the school and university communities, all organized around cultural-historical activity theories of learning and development; promote learning among undergraduate and elementary school children. The authors concluded that the way individuals interpret and articulate problems, the mediation strategies they use, and how they define and negotiate their roles and responsibilities for knowledge production relate to their local community and its history of practices. Jurdak (2006) contrasted theoretically and empirically the problem solving of situated problems in school and the real world. Thirty-one last year high school students in the scientific stream solved three potentially experiential problem tasks. The results indicated that there are fundamental identifiable differences among the activities and the activity systems of problem solving in the real world, situated, and school contexts. Jurdak and Shahin (2001) compared and contrasted the nature of spatial reasoning by practitioners (plumbers) in the workplace and students in the school setting while constructing solids with given specifications, from plane surfaces. The results of this study confirmed the power of activity theory and its methodology in identifying and explaining differences between the two activities in the two different cultural settings.

In general, sociocultural mathematics education research focused mainly on understanding the learning of mathematics in different sociocultural contexts. However, a close examination of the details of these studies show a little engagement with the role of power in influencing the learning of mathematics in different cultural settings.

3 Bourdieu's Theory and the Political Dimension of Mathematics Education

The ICME-13 Topical Survey on *Social and Political Dimensions of Mathematics Education, Current Thinking* (Jurdak et al. 2016) identified a number of factors that may account for inequities in access and distribution of mathematics education. Socio-economic include socioeconomic status, ethnicity, gender, material conditions within which mathematics education takes places, and nature of a society's economic structure. Mathematical factors include the nature of mathematics as a discipline and mathematics as a regime of truth. Ideological factors include ideologies/philosophical underpinnings underlying state policies and practices of actors in mathematics education. One thesis of this chapter is that the power construct in Bourdieu's field theory may provide an explanatory framework to account for inequities and student marginalization in mathematics education.

3.1 Bourdieu's Construct of Power

Bourdieu's construct of power rests on a complex interplay of the concepts of field, habitus, and capital. A *field* is a social network or configuration, which has an object and exists in specific location. It consists of a structured space of positions and a space of positions-takings (Bourdieu 1993). According to Bourdieu, a field is not only a social field but also a *field of power* since the state of the relations between positions in the field is the result of agents striving—intentionally and unintentionally—for goods and resources (i.e. capital) that are specific to the field. *Habitus* is a complex interplay between *individual internalization of past socializations and those of present* (Bourdieu 1990). A social agent, not only acts on current circumstances, but also internalizes them to become another layer to add to those from earlier socializations. Habitus undergoes continual restructuring. *Cultural capital* is the product of education and refers to forms of knowledge, skills, education and academic credentials, etc. *Social capital* refers to resources based on group membership. *Economic capital refers* to material wealth and time, which can be cashed in any part of society. *Symbolic capital* is nothing other than capital, in whatever form, when perceived by an agent, without questioning the basis of its existence and basis of that capital, as evident and legitimate.

The essence of Bourdieu's construct of power is exchanging capital from one form to another and disguising that in the form of symbolic capital. For example, those who have the advantage to exchange and disguise their possession of economic capital, and subsequently cultural or social capital, in the form of symbolic capital, that is honor or prestige, assume the power to dictate systems of meaning on those who do not have that advantage. Applied to mathematics education, Bourdieu's construct of power proposes that advantaged social groups have the opportunity to exchange and disguise their possession of economic, cultural or social capital, in the form of symbolic capital (honor or prestige), assume the power to dictate policies, practices, and beliefs on mathematics education institutions and actors. Bourdieu's construct of power may account for the inequity and marginalization in mathematics education attributed to socioeconomic factors. Economically advantaged social groups may exchange and disguise their possession of economic capital to impact policies (curriculum, for example), practices (use of technology), and beliefs (student self-concept). Socially advantaged social groups such as male-dominated or ethnic majority dominated social groups may similarly exchange and disguise their possession of social and or cultural capital to impact policies (racial-based school compositions), practices (gender discrimination), and beliefs (student alienation). The state may exchange and disguise its possession (sanctioned by legitimate popular mandate) of economic, social, and cultural capital, to enact ideologically motivated policies that may disadvantage certain social groups. Mathematics educators may exchange and disguise their possession of their cultural capital (knowledge of mathematics and its pedagogy) to impose mathematical meanings that may exclude certain groups of students.

3.2 Mathematic Education Research Using Bourdieu's Theory

Recently, a few research studies have addressed the role of power in mathematics education, using Bourdieu's filed theory. All these studies focused on the power rather than the sociocultural dimension. For example, Jorgensen et al. (2014) examined two children's mathematical learning trajectories to highlight how school mathematics practices allow greater or lesser access to school mathematics depending on the cultural backgrounds and dispositions of the learners. Their findings were consistent with Bourdieu's original field theory. Nolan (2016), using Bourdieu's social field theory, explored discourses of school mathematics class-rooms as experienced by two novice secondary mathematics teachers. The data reveal that the ways in which the two novice mathematics teachers carefully negotiate space for enacting agency amid school social structures, are consistent with Bourdieu's social field theory in that the social structures of a field both "constrain and (re)produce the becoming teacher" (p. 328). However, Nolan noted that the letters from the field written by the two novice teachers during the study provided discourses which competed with the discourses offered by their teacher educators/researchers in their teacher education program.

In general, Bourdieu-inspired mathematics education research attempted to establish that even mathematics education is not immune from the impact of power that enable advantaged groups to use their possession of capital to reproduce existing inequities in mathematics learning. However, these studies show little engagement with the role of sociocultural factors in shaping mathematics learning.

4 A Locally Attuned Version of Bourdieu's Field Theory and Mathematics Education

On its surface, Bourdieu's field theory may lead to the pessimistic conclusion that education is bound to reproduce existing inequities because it considers culture a carrier of capital through the accumulated internalized socializations of the habitus of the social agent. Ferrare and Apple (2015) suggest that Bourdieu's field theory needs not arrive at that conclusion and propose an elaboration of Bourdieu's field theory to be more attentive to understanding of how actors construct experience and struggle over meanings in local contexts such as individual schools and universities.

In the following paragraphs, I first introduce Ferrare and Apple's (2015) locally attuned version of Bourdieu's field theory followed by my interpretation of this theory as it applies to mathematics education.

Ferrare and Apple (2015) argue that "the most important problem inherited from Bourdieu's field theory stems from his disinclination—shared by many in sociology —to venture into the realm of individual perception and experience" (p. 45). They argue that:

> Bourdieu's primary emphasis on the macro view of cultural fields obscures an under-standing of how educational actors directly experience and make sense of the pedagogic qualities—what we will later call 'affordances'–inherent in local field positions, practices and meanings.
>
> (p. 45)

To build upon the problem inherited from Bourdieu's field theory, Ferrare and Apple have drawn upon insights from social psychological field theory, ecological psychology and in relational sociology.

> The first insight suggests the need to consider a greater degree of complexity in the mental structures constituting the individual (or group) life space. Our second point makes the case for extending the affective dimension of field theory to include the values that inhere in objects, not just an individual's habitus. Finally, we suggested that it is important to consider fields from a phenomenological perspective, which means that we must be attentive to the local institutional positions in which actors experience their day-to-day lives. (p. 52)

According to Ferrare and Apple (2015), Bourdieu constructed his version of field theory "in which social actors experience fields as both arenas of force and arenas of struggle". (p. 48). They note that Bourdieu emphasized and developed fields more as arenas of force than that of struggle. Social actors experience fields as an arena of force in the sense that fields are rules that "direct normative values, regulate actions, reward ontological complicity, and place sanctions on transgressors" (p. 48). Social actors experience fields as an arena of struggle in the sense that actors who possess "an advanced feel for the game (i.e. a *habitus attuned to the situation*), have more opportunities to adapt and improvise strategies to achieve success in the field. It is this feel for the game that enables some actors the freedom to know when to take risks—to engage in subversion strategies—and when to 'dig in' and fight to con-serve the present rules of engagement" (p. 48) [*italics are mine*].

Based on these insights, Ferrare and Apple (2015) proposed a locally attuned version of field theory that extends the concept of habitus to better account for the information that inheres in local field positions in educational contexts as well as recognizes how students perceive this information as constraints and affordances related to their educational experiences, goals and aspirations. This locally attuned version "shifts the deficit from the individual to the field itself, and provides a space to interrupt the arbitrary and inequitable valuation of certain cultural forms over others" (p. 54).

The implications of the locally attuned version of Bourdieu's field theory to school mathematics are far-reaching. Historically, the dominant public view of school mathematics is that mathematical ability is genetically endowed or culturally inherited. The locally attuned theory challenges that view by enabling the habitus of the individual student to interrupt the complicity between the advantaged groups and school mathematics practices. This theory shifts the deficit in mathematics learning from the individual student to what is lacking in terms of democratization of school mathematics education. By doing so, students are provided with more opportunities to best use their own perceptions of the local constraints and affordances related to their experiences, goals and aspirations to adapt and improvise strategies to achieve success in mathematics learning.

5 Modular Integration of the Locally Attuned Version of Bourdieu's Theory and Engeström's Activity System

This is not the first time that CHAT's tradition and Bourdieu's field theory are proposed to be combined in one theory. Williams (2012) proposed to extend CHAT to incorporate Bourdieu's sociology in order to bring together theories in "a joint theory of education as both development and re-production of labour power, in which use and exchange value both have their place (in commodity production)" (p. 57). This extension can incorporate the "CHAT perspective on the 'cultural development of the mind' to "the use of mathematics as a tool for the critical, scientific examination of society" (p. 70). This chapter proposes to integrate Bourdieu's field theory, as extended by locally attuned version of Bourdieu's field theory (Ferrare and Apple 2015) and Engeström's activity system, in order to seek a common foundation to sociopolitical mathematics education.

The modular integration connects existing theories without replacing them or subsuming one within the other. "The original theories are treated as modules, pulled off the shelf and plugged into one another as needed, and still available in their un-integrated form for use in other integrations. This is deeply analogous to the integrated circuit in electronics" (Markovsky et al. 2008, p. 347).

5.1 The Two Theories as Candidates of Modular Integration

At a practical level, modular integration means that two integrated theories, plugged into one another, are able to provide better explanations than they could individually. According to Kalkhoff et al. (2010), to be candidates for modular integration, theories should be coherent and clear, empirically supported, and share concepts and constructs by which they can be integrated. Bourdieu's field theory and its locally attuned version (Ferrare and Apple 2015) as well as Engeström's activity system are well-established theories as explained earlier. These two theories are empirically supported in the area of mathematics education (see Sects. 2.4 and 3.2). Next, we explore the common concepts and constructs between the two theories.

Both CHAT and Bourdieu's locally attuned version share the concepts of *culture* and *history*. CHAT assumes that the understanding of human activity can only be within the communal collective meanings of the cultural context in which the activity is enacted. In Bourdieu's field theory, culture plays a pivotal role by being a carrier of capital through the internalized socializations of the habitus of the social agent. CHAT and Bourdieu's field theory posit that the *historical* dimension is an indispensable ingredient of their theoretical foundation. According to Roth et al. (2012), CHAT assumes that human activity occurs in a historical context in the sense that the historical context contextualizes the activity itself. Therefore, what we are observing today is different from what we might have observed in the same context ten years earlier. According to Bourdieu (1990), habitus refers to individual history. Habitus undergoes continual restructuring. The habitus acquired in the family is at the basis of the structuring of school experiences and the habitus acquired in school is in turn at the basis of all subsequent experiences. Bourdieu's focus on the agency of the social actors, whose habitus is more attuned to the situation in their local contexts, was obscured by his tendency to focus more on the structure of macro-level field positions. Ferrare and Apple (2015) contribution is in their attempt to bring into prominence an under-developed concept of Bourdieu i.e. the nuanced "understanding of how actors construct, experience and struggle over meanings in local contexts such as individual schools and universities" (p. 45).

5.2 Applying Modular Integration to the Two Theories

According to Markovsky et al. (2008), the "connecting threads link theoretical elements" (p. 348) hold together theories that form the fabric of knowledge. Some of those threads are terms, definitions, propositions, arguments, and scope conditions. The two theories under consideration have many terms in common such as culture and history. However, the definitions of those terms are different in the two theories and hence are not good candidates to be 'connecting threads'. On the other hand, 'agency of the social actor' and 'local contexts' are two terms which have similar definitions in both theories. In activity theory, the choice of actions as well as the artifacts is contingent on the learner's choices and consciousness of the potential effectiveness of the actions in realizing the intended goal. Moreover, local temporality is involved in an activity (Roth et al. 2012). In the locally attuned version of Bourdieu's theory, social actors experience fields as an arena of struggle in the sense that actors who possess "an advanced feel for the game (i.e. a habitus attuned to the situation), have more opportunities to adapt and improvise strategies to achieve success in the field (p. 48).

In summary the agency of the social actor and the local contexts are the two connecting threads that bring the two theories together. The integrated theory may account for both the social and political dimensions of mathematics education.

Specifically, on the one hand, it provides a lens that enable mathematics educators to see 'both sides of the coin' of sociopolitical mathematics education: The integrated theory supports the view of school mathematics as both a social cultural activity system as well as a field of force and of struggle. On the other hand, the integrated theory provides a perspective that may open a window to avoid the deadlock that arises from the assumption that exclusion in school mathematics is something inherent to the school system and to end school mathematics, as we know it (Pais 2012). Because it shifts attention "toward the qualities that inhere directly in school-level social structures, practices and meanings", the locally attuned version of Bourdieu's field theory (Ferrare and Apple 2015) shifts the deficit from the individual to the field itself, and provides a space for the social actor to interrupt exclusion and marginalization in the local context.

6 Implications of the Integrated Theory

6.1 Implication for Mathematics Education Research

One theoretical implication of the proposed integrated theory is the shift in the focus of mathematics education research. The locally attuned module (Ferrare and Apple 2015) in the integrated theory calls for focusing on the structures of practice and meaning of mathematics education in local educational contexts (e.g. schools), in order to understand how "variations in local social configurations can meaningfully shape the ways that students perceive and interpret constraints and affordances that inhere in educational settings" (p. 53). This focus is not alien to mathematics education research. In fact, this kind of focus appeared in two highly-cited studies done at the beginning of this century by Vithal (2003) in South Africa and by Gutstein (2006) in the United States. Vithal investigated what happened in a mathematics classroom in local contexts when student teachers attempted to use a social, cultural, and political approach which integrates a critical perspective to the school mathematics curriculum in post-apartheid South Africa. Gustein's pedagogical goal was to create conditions for students to develop agency in a middle-school mathematics classroom in a Chicago public school in a Latino community. Both studies reported some short-term gains and a number of challenges embedded in the system itself. Both studies used the social and political dimensions in their studies. What is surprising is that both studies aimed at what Ferrare and Apple (2015) called for, i.e. understanding how the local institutional social configurations "shape the ways that students perceive and interpret constraints and affordances that inhere in the educational settings" (p. 53). Unfortunately, this line of research did not continue for a variety of reasons. Based on the proposed integrated theory, I support and call for reviving sociopolitical mathematics education research that focuses on the structures of practice and meaning of school mathematics education in local educational contexts in order to understand how students perceive the constraints and affordances in those contexts. The second theoretical implication is the need to have a research paradigm whose purpose is to understand how social school structures shape students' interpretation of mathematics education in terms of the constraints and affordances inherent in local contexts. The research paradigm that best serves this purpose would include a thick understanding of experience and meaning of the social actors. One challenge for the integrated theory is its ability to account for phenomena that its modular components cannot adequately explain. One such phenomenon is the many mathematics educators who were able to achieve academic success (the author is one of them) despite the fact that they were disadvantaged in terms of economic, social, or cultural capital. Bourdieu's field theory would not be able to account for such a phenomenon. Again, activity theory has little to contribute in this regard. From the perspective of the integrated theory, this phenomenon is amenable to be studied through case studies of such disadvantaged individuals in terms of their interpretation of the constraints and affordances in their local educational contexts to beat the system as a field of force.

6.2 Implications for School Mathematics Practices

An implication of the integrated theory is that it politicizes the mathematical practices in the activity system of school mathematics. The concept of the division of labor in the activity system is neutral in the activity system. From a political perspective, division of labor is essentially a political act in terms who and how the division of labor is done. In school mathematics classroom, division of labor practices are normally characterized by teacher domination on the pretext that the teacher possesses, compared to students, superior knowledge of mathematics (cultural capital), and hence has the option to impose own meanings as legitimate and necessary for student success. A division of labor in which the role of teacher moves away from a definer of mathematical meanings to an arbiter of student meanings, enable students to experience and make sense of mathematics as it relates to their experiences may interrupt the imposition of meanings by teacher. Another set of practices in activity theory relate to access and distribution of *mediating artifacts*. The activity theory is silent on the role of power differential associated with the inequities that might arise from differential access and distribution of 'mediating artifacts' i.e. the material and symbolic tools for learning and teaching school mathematics. For example, in schools, which use digital technologies for teaching and learning mathematics, students from higher socioeconomic families, compared to students from lower socioeconomic families, are more likely to have more exposure and higher technological literacy in using those tools. Student agency to interrupt inequities in this case may not be enough to interrupt the existing inequities, and consequently the school itself may have to be more responsive and adaptive to the needs of disadvantaged students. *Symbolic mediating artifacts* play a subtle role as possible instruments of exclusion and marginalization in the learning and teaching of mathematics. For example, competencies in language and mathematics, which mediate the learning and teaching school mathematics, are culturally constituted and ingrained in the habitus, and therefore, student agency would not be in a favorable position to interrupt the inequities arising from the complicity between the language and mathematical practices in school and those of the dominant groups. In conclusion, the inequities arising from the differential access and distribution of the practices espoused by activity theory are political in nature and hence need to be addressed in the political arena.

6.3   Implications for School Mathematics Policies

One implication of the integrated framework is that the interruption of inequities in mathematics education at the school level requires favorable policies and practices at the system level. According to Jurdak (2011), from an activity theoretic perspective, mathematics education at the national level is a complex nested hierarchical 3-layer activity system: The classroom, school, and the state system. Because each system nests within the next higher one, the societal relationships of power of a higher system carry over to the lower systems and eventually to the student at classroom level. Thus, the existence of favorable policies and practices at the system level is necessary for the successful implementation of the integrated theory at the school level. The mathematics curriculum is an example of a system policy that might constrain the ability to interrupt inequality because of its structure and orientation. A mathematics curriculum that does not provide enough space for students to perceive mathematics learning, enacted in the locality of the school, as an affordance related to their educational experiences, goals and aspirations, would not be favorable to enabling students make sense of the opportunities in local positions, practices, and meanings. The proposed integrated theory shifts attention toward the qualities that inhere directly in school-level social structures, practices and meanings. This shifts focus from asking what students are lacking to what is lacking from the social structures and cultural models of schools. Thus, the practical goal of attempting to democratize access and distribution of school mathematics education leads to one of "democratizing the very construction of the entire space of positions and cultural forms constituting schools (Ferrare and Apple 2015, p. 46). As researchers and mathematics educators, we have little or no leverage when it comes to policy-making. However, we have the ethical responsibility to struggle for restructuring and democratization of the social structures of our school systems.

7 A Personal Narrative on My Theoretical Journey to Sociopolitical Mathematics Education

In recent years, it has been my conviction and practice to illuminate my presentation of theoretical argumentation by a personal narrative on why and how my experience shaped those ideas and argumentation (Jurdak 2016a). Thus, I conclude this chapter with a personal narrative to tell the story of why and how the idea of linking the theories of Engeström, Bourdieu, and Apple came into being.

At the turn of the last century, while I was engaged in a research project on comparing and contrasting mathematical problem solving in and outside school, I came across activity theory as a possible explanatory model for my research. Activity theory proved to be a powerful theory to explain the differences in problem solving in school and the real world in terms of the differences in their sociocultural contexts. I was so fascinated by activity theory to the point that, for some years, I used it as a lens to make sense of mathematics education (and even all educational) practices.

In 2007, my colleague Saouma BouJaoude, a science educator, and I obtained a grant to start a school-based reform project called TAMAM, an acronym derived from the Arabic title of the project which consists of the initials of "school-based reform" in Arabic (al-Tatweer Al-Mustanid ila Al-Madrasa) and which means "perfect" in colloquial Arabic (address: http://tamamproject.org/). The project aimed to develop a school-based grounded theory of educational reform in the Arab region that would provide policymakers with research-based recommendations for implementing educational reform in their countries. It involved a partnership between the American University of Beirut and nine school teams from three Arab countries. The project included a variety of mediating artifacts such as conferences, action research school-based projects, and reflective practice.

From the beginning, I could conceive of TAMAM as an activity system but, as the project proceeded, I felt that this framework did not capture the complexity of learning that was taking place. Unlike my university students, the TAMAM participants brought with them to their learning in TAMAM, a rich capital of actual experiences, which in many respects exceeded my experiences as a university professor. At this point, I started to believe that activity theory was not enough to explain TAMAM learning mostly based on reflective practice. Freire's emancipatory ideas came very handy as a powerful tool to explain this kind of complex learning. This was when the narrative came to my mind, and I decided to transform the individual experiences of TAMAM participants into stories written by the individuals themselves and in their own native language (Arabic), which was published as an e-book entitled '*School-based reform (TAMAM): Voices from the field (in Arabic)*' (Jurdak 2016b). In retrospect, my encounter with Freire's ideas was my first engagement with the 'soft' political aspects of mathematics education. However, the professional development, which helped transform TAMAM participants, failed to 'reform' their respective schools. Neither activity theory nor Freire's emancipatory education could provide a convincing explanation for TAMAM success in the 'development' of the participants and its failure in achieving the institutional 'reform' effort. This discrepancy led me to Bourdieu cultural reproduction theory.

Although different, the nine schools have one thing in common i.e. each school has achieved, through its sociocultural history, a 'status' earned through accumulating cultural, social, and economic capital. Schools did not seem to be ready to change (reform) their practices because they wanted to protect their privileged status. Schools maintain their power by reproducing the culture that produced them. However, it was difficult for me to understand the development of the TAMAM participants and the resistance of their institutions without engaging both activity theory and Bourdieu's cultural reproduction theory. For me then, the development and the resistance were two sides of the same coin. That was the genesis of integrating the two theories. However, I faced a considerable difficulty in finding common linking threads between the two theories, which led me to Ferrare and Apple. That was my frame of thinking when I decided to contribute to this book.

References

Bourdieu, P. (1990). *The logic of practice*. Stanford, CA: Stanford University Press.

Bourdieu, P. (1993). *The field of cultural production: Essays on art and literature*. Cambridge: Polity Press.

Bourdieu, P., & Passeron, J. C. (1990). *Reproduction in education, society and culture* (2nd ed.). London: Sage Publications.

Cobb, P. (2006). Mathematics learning as a social practice. In J. Maasz & W. Schloeglmann (Eds.), *New mathematics education research and practice* (pp. 147–152). Rotterdam: Sense Publishers.

Engeström, Y. (2015). *Learning by expanding: An activity-theoretical approach to developmental research (second edition)*. New York: Cambridge University Press.

Engeström, Y. (2001). Expansive learning at work: Toward an activity theoretical reconceptualization. *Journal of Education and Work, 14*(1), 133–156.

Ferrare, J., & Apple, M. (2015). Field theory and educational practice: Bourdieu and the pedagogic qualities of local field positions in educational contexts. *Cambridge Journal of Education, 45* (1), 43–59.

Freire, P. (1970/2013). *Pedagogy of the oppressed*. New York: Bloomsbury.

Gutiérrez, R. (2013). The socio political turn in mathematics education. *Journal for Research in Mathematics Education, 44*(1), 37–68.

Gutstein, E. (2006). *Reading and writing the world with mathematics: Toward pedagogy for social justice*. New York: Routledge.

Jorgensen, R., Gates, P., & Vanessa Roper, V. (2014). Structural exclusion through school mathematics: Using Bourdieu to understand mathematics as a social practice. *Educational Studies in Mathematics, 87*, 221–239.

Jurdak, M. (2006). Contrasting perspectives and performance of high school students on problem solving in real world, situated, and school contexts. *Educational Studies in Mathematics, 63*(3), 283–301.

Jurdak, M. (2011). Equity in quality mathematics education: A global perspective. In B. Atweh, M. Graven, & W. Secada (Eds.), *Mapping equity and quality in mathematics education* (pp. 131–144). Dordrecht: Springer.

Jurdak, M. (2016a). *Learning and teaching real world problem solving in school mathematics—A multiple-perspective framework for crossing the boundary*. Switzerland: Springer.

Jurdak, M. (Ed.). (2016b). *School-based reform (TAMAM): Voices from the field (in Arabic)*. Beirut: Arab Thought Foundation. Available at: http://tamamproject.org/research/tamam-voices-from-the-field/.

Jurdak, M., & Shahin, I. (2001). Problem solving activity in the workplace and the school: The case of constructing solids. *Educational Studies in Mathematics, 47*(3), 297–315.

Jurdak, M., Vithal, R., de Freitas, E., Gates, P., & Kollosche, K. (2016). *Social and political dimensions of mathematics education, ICME-13 topical surveys*. New York: Springer. doi: https://doi.org/10.1007/978-3-319-29655-5_1. Address on line: http://link.springer.com/book/10.1007/978-3-319-29655-5.

Kalkhoff, W., Friedkin, N. E., & Johnsen, E. C. (2010). Status, networks, and opinions: A modular integration of two theories. In *Advances in group processes* (pp. 1–38). Bingley: Emerald Group Publishing Limited.

Leont'ev, A. N. (1978). *Activity, consciousness, and personality*. Englewood Cliffs: Prentice-Hall.

Leont'ev, A. N. (1981). The problem of activity in psychology. In J. Wertsch (Ed.), *The concept of activity in Soviet psychology* (pp. 37–71). Armonk, NY: M. E. Sharpe.

Lerman, S. (2000). The social turn in mathematics education research. In J. Boaler (Ed.) *Multiple perspectives on mathematics teaching and learning* (pp. 19–44). Westport, CT: Ablex.

Lerman, S. (2006). Cultural psychology, anthropology and sociology: The developing 'strong' social turn. In J. Maasz & W. Schloeglmann (Eds.), *New mathematics education research and practice* (pp. 171–188). Rotterdam: Sense Publishers.

Lucariello, J. (1995). Mind, culture, person: Elements in a cultural psychology. *Human Development, 38*(1), 2–18.

Markovsky, B., Dilks, L. M., Koch, P., McDonough, S., Triplett, J., & Velasquez, L. (2008). Modularizing and integrating theories of justice. In *Justice* (pp. 345–371). Bingley: Emerald Group Publishing Limited.

Nolan, K. (2016). Schooling novice mathematics teachers on structures and strategies: a Bourdieuian perspective on the role of "others" in classroom practices. *Educational Studies in Mathematics, 92*, 315–329.

Pais, A. (2012). A critical approach to equity. In O. Skovsmose & B. Greer (Eds.), *Opening the cage* (pp. 49–91). Netherlands: Sense Publishers.

Roth, W., Radford, L., & La Croix, L. (2012). Working with cultural-historical activity theory. *Forum: Qualitative Social Research, 13*(2), 1–20.

Shulman, L. S. (1970). Psychology and mathematics education. In E. Begle (Ed.), *Mathematics education* (pp. 23–71). Chicago: National Society for the Study of Education.

Skovsmose, O. (2011). *An invitation to critical mathematics education*. Rotterdam: Sense Publishers.

Skovsmose, O., & Greer, B. (2012). *Opening the cage: Critique and politics of mathematics education*. Rotterdam: Sense Publishers.

Stone, L. D., & Gutiérrez, K. D. (2007). Problem articulation and the processes of assistance: An activity theoretic view of mediation in game play. *International Journal of Educational Research, 46*(1), 43–56.

Vithal, R. (2003). *In search of a pedagogy of conflict and dialogue for mathematics education*. Dordrecht, The Netherlands: Kluwer Academic Publishers.

Vygotsky, L. S. (1978). *Mind in society*. Cambridge, MA: Harvard University.

Williams, J. (2012). Use and exchange value in mathematics education: Contemporary CHAT meets Bourdieu's sociology. *Educational Studies in Mathematics, 80*(1/2), 57–72.

Towards an Ethics of Mathematical Application

Felix Lensing and Hauke Straehler-Pohl

Abstract In the light of growing public attention to the influence of algorithms on our lives, this chapter addresses the question of how an ethical perspective on mathematical application could be conceptualised in contemporary late-modern societies. Firstly, we recapitulate some of the recent theoretical developments on the ethics of mathematical application in the field of mathematics education (Skovsmose in Mathematics education in a knowledge market: developing functional and critical competencies. Opening the research text: insights and in(ter)ventions into mathematics education. Springer, New York, pp. 159–188, 2008; de Freitas in Int Electr J Math Educ 3(2):79–95, 2008). Secondly, based on the work of the sociologist Luhmann (Thesis Eleven 29(1):82–94, 1991), we develop theoretical outlines of an ethics of mathematical application as a reflexive theory of moral communication on mathematical application. We then move into the sphere of the social and confront these theoretical considerations with a critique of the ideology of "solutionism". Solutionism refers to a semantics that links the mathematisation of the social to 'the morally good'. This critique leads us to suggest firstly, developing an ideology critique of the underlying semantics as a desideratum; and secondly, a systematic further development of an ethics of mathematical application that could inform moral communication on mathematical application in (critical) mathematics education.

Keywords Mathematical application · Ethics · Mathematization of social process Critical mathematics education

F. Lensing (✉) · H. Straehler-Pohl
Freie Universität Berlin, Berlin, Germany
e-mail: lensinfe@zedat.fu-berlin.de

H. Straehler-Pohl
e-mail: h.straehler-pohl@fu-berlin.de

© Springer International Publishing AG 2018   35
M. Jurdak and R. Vithal (eds.), *Sociopolitical Dimensions of Mathematics Education*, ICME-13 Monographs, https://doi.org/10.1007/978-3-319-72610-6_3

1   Introduction

Today, it seems to be common sense that 'mathematics is everywhere'. This is not a mere slogan anymore that is solely promoted by mathematics educators to proclaim the relevance of their subject. The number of mathematisations that are 'colonising' the every-day world is "growing exponentially" (Ernest 2001, p. 287). Likewise, the number of reflections which discuss the consequences of this development (most commonly spread by the mass media) is at an all-time high. To name just one example, the article "How algorithms rule the world", published in *The Guardian*, draws on insights from the recent NSA revelations: "The NSA revelations highlight the role sophisticated algorithms play in sifting through masses of data. But more surprising is their widespread use in our everyday lives. So should we be more wary of their power?".[1] In many cases, mathematics appears as a power of its own, changing the world according to its supposed 'own will' and thereby re-programmes the conditions of our lives beneath our consciousness (Han 2017). However, the myth that 'mathematics is everywhere' is contingent upon the fact that humans *apply* mathematics to all spheres of life, including the social sphere.

If we accept the thesis that the social process of mathematisation increasingly influences all different aspects of our lives, reflection on the conditions and consequences of mathematisations becomes more important than ever. Then, one task for mathematics educators and researchers is to draft their possible contributions to the discussion on the "formatting power" (Skovsmose 1994, p. 43) of mathematics. As always, the first step would be to pose a good question. In any case, the evolving forms of interaction and communication will dramatically change the ways in which we see the world and ourselves. This is why the question at stake cannot be about *if* we want mathematisations to regulate the social spheres of life. Instead, it needs to be *how* we can develop a critical stance that allows us to reflexively deal with the mathematisations that in turn shape our lives. Thus, any form of critique that goes beyond a simple rejection of the social process of mathematisation has to rely on a reflexive theory that allows us to confront "what is the case with what is not the case but could become the case" (Skovsmose and Borba 2004, p. 214). Moreover, such a theory would need to exercise this critique from the inside of the object under investigation. In mathematics education theory, that is the point where we enter the domain of critical mathematics education since "reflection is a characteristic of being critical" (Skovsmose 2008, p. 159). In practice, critical mathematics education aims to initiate teaching and learning processes that allow students to turn mathematics against itself. That is, students are encouraged to reflect upon the philosophical grounds of mathematics as well as the conditions and consequences of mathematical applications. Such reflections can reveal political and ethical concerns that are usually not addressed in mathematics classrooms. This chapter shall be read as a contribution to the ongoing endeavour to identify ways to

---

[1]Retrieved from: http://www.theguardian.com/science/2013/jul/01/how-algorithms-rule-world-nsa on September 8th, 2015.

empower students to become critical citizens. Thereby, we focus on the question of how an ethics of mathematical application, in light of the all encompassing process of mathematisation, could look like. By seeing the mathematisation of the social as a major challenge to both mathematics education research as well as the mathematics education practice, we will firstly recapitulate some of the recent theoretical approaches developed on the ethics of mathematical application (de Freitas 2008; Skovsmose 2008). Secondly, we aim to sketch out the theoretical outlines of an ethics of mathematical application as a reflexive theory of moral communication on mathematical application. Thirdly, we will move into the sphere of the social and use our theoretical work to exemplarily analyse a social phenomenon: The phenomenon that the mathematisation of the social and 'the morally good' seem to enter a peculiar bond in the semantics[2] propagated by the ideological leaders of digitalisation (such as Facebook, Google, Amazon, and the like). Finally, we suggest: (1) the development of an ideology critique of the underlying semantics as a desideratum, and (2) a further development of our sketched framework of an ethics of mathematical application that could inform moral communication on mathematical application in (critical) mathematics education.

2 Ethical Filtration: What Could 'Being Critical' Mean?

In order to better understand the potential effects of the application of mathematics, Skovsmose (2008) investigates "mathematics in action" (p. 163). Therefore, he observers what people do when they mathematise a social practice. With the focus on how mathematics is brought into action for the organisation of the practices, he identifies a phenomenon that he calls "ethical filtration" (ibid.). As soon as a social practice is abstracted to numbers, variables, and the relations between them, all considerations with direct reference to the practice seem to vanish. Immediately, the focus solely resides on the accuracy of the accompanying transformations and calculations. Consequently, this means that the process of mathematisation loses its contact to the concrete situation. Thereby, the model seems to become blind to its own origin in the 'real' social situation (and so does the modeling agent). In other words, the process of mathematisation tends to entail a moment where contingency, subjectivity, and materiality are stripped away (de Freitas 2008) and the (supposedly) immanent logic of mathematics takes over. Nonetheless, ethical filtration itself is not to be understood as a malicious or imprudent (mis)use of mathematics in applications. Instead, it turns out to be "a general feature of bringing mathematics

---

[2]We conceptualise a semantics as a self-description of the society *in the society* that is articulated in communication. When societies change, so do the available possibilities for the members of societies to communicate about the society they live in, in other words, semantics change. Simultaneously, when semantics change, so do the societies that make use of them. A semantics is thus a pre-condition that shapes communication. In turn, it is shaped by all communications that are recursively producing the society.

into action" (p. 167). Thus, it should be considered as a phenomenon that is *immanent to the process of mathematisation itself*. Due to findings like these, one of the overall aims of critical mathematics education is precisely to counteract such exercising of mathematics without any accompanying reflections. In regard to this, Skovsmose (2008) poses an important question: "What does it mean to establish an ethical perspective on mathematics in action?" (p. 166).

Skovsmose (2008) addresses this question in his report on his project "Family support in a Micro Society". The project aimed to make students "experience how mathematics can be brought into action—how it may be part of a decision making process and, in this way, becomes part of peoples' reality" (p. 166). The participating students were divided into different groups. Each group was assigned to a fictional micro society consisting of 24 families that were further described in essays:

> Each group had to formulate principles according to how they wanted to distribute child benefits among families. The amount of money available was given, but each group could formulate any criteria according to how they wanted to distribute it. Next they had to provide an algorithm for distributing the money. [...] In the process of turning the verbally-formulated principles for distribution into functional algorithms, the students experienced how the original principles needed to be simplified. At times the principles were almost ignored when mathematics was brought into operation to do the distribution. The students experienced the general phenomenon that when mathematics is brought into action, a new discourse takes over. (ibid.)

However, the observation that the students "experienced" the phenomenon of ethical filtration should not mislead us about the critical effects that this experience actually unfolded. That is, the experience of ethical filtration does not automatically lead to the development of a critical stance towards the exercised mathematical models in particular and mathematical application in general. It rather opens up a space of possibility that *might* "indicate what 'being critical' could mean in educational practice" (Skovsmose 2008 p. 167). In other words, *there is always a gap between the potentiality of a critical stance evolving from the experience of ethical filtration in the mathematical modelling process and the actuality of a critical stance that is yet to be developed*. This view is supported by the detailed consideration of some students' utterances from the project in Skovsmose's (1994) dissertation. Not seldom with a touch of irony, he depicts how the students are rather "absorbed in the technical task of making the distribution" (p. 138) than being critical to it. For example, one student reflects on his action in an early unit: "I see, the age of the child is missing. Anyway, the family lives in number 13, so the age may as well be 13 too!" (p. 127). Another student, reporting on a late unit where the teacher intentionally initiated a discussion on the differences of the models and their sociopolitical implications, states "We agree that a difference [between different distribution models] exists, but anyway we have made the calculations correctly" (p. 128).

The reflections on the project can be considered as evidence for Lundin's (2012) assumption that the practice of school mathematics is often informed by an ethics that "establishes mathematical knowledge as good, by making such knowledge have beneficial consequences" (p. 81). The reflections on the project can be considered as evidence. In general, this effect can certainly be recorded for any school subject; however, the unique characteristics of school mathematics goes one step further. Mathematical knowledge is not only identified as morally good, but at the same time *allegedly 'true' mathematical knowledge is established as always 'out of reach'*. That means that students learn to privilege mathematical knowledge by experiencing the beneficial consequences of applying it. They further learn that the mathematics they apply is always just an impure form - a form of mathematics being distorted by the imperfection of school. However, this also presupposes that an application of 'true' mathematical knowledge, which would solve problems in a true and not solely in an approximate manner, is principally possible as long as the limitations of mathematics in action are solely attributed to the limited complexity of school mathematics to fit the complexity of the 'real' world. In other words, students may experience mathematics as a "pure [...] and wholly logical knowledge, which [...] happens to be useful because of its universal validity" (Ernest 2001, p. 279) not despite, but precisely because of the experience that "the original principles needed to be simplified" (see above) in order to bring mathematics in action within the wider frame of school mathematics, which rewards mathematisation as an end in itself. Instead of leading to a critical stance towards mathematical application, such experience may just as well reinforce an absolutist conception of mathematics.

But, the claim that mathematics represents an eternal body of knowledge which can be stripped of any contingency was also heavily questioned within the mathematical discourse in the beginning of the 20th century: Gödel (1931) showed that it is always possible to construct theorems that are *undecidable from the inside of a formal mathematical system*. In other words, mathematics cannot bootstrap its own conditions of possibility. Thus, since it depends in its very constitution on an extra-mathematical, subjective act that cannot be grounded in mathematics itself, mathematics has to be understood as radically political. Therefore, any ethics that advises us to simply identify mathematics as morally good is actually a "quasi-ethics" because it can only justify the superiority of mathematics as the form of seeing the world by implicitly presupposing an ontological unity between mathematics and being that has become more than questionable. Following this line of thought, we are thrown back into the gap between the potentiality of a critical stance and its yet to be developed actuality where any 'struggle' for a critical stance towards mathematics and its applications seems to take place. In order to unfold its critical potential, an activity like the "Family support in a Micro Society" would thus need to somehow conduct a juggling act: The activity needs to avoid a one-sided moral communication on mathematics (mathematics is morally good in its very structure), while simultaneously, bearing in mind that it is also the initiation of reflective processes themselves that could implicitly reinforce the bond between 'the morally good' and the application of mathematics. Therefore, as a first approximation, we suggest that it is important to: (a) pay attention to not credit students in case that they relativise their models (e.g. "we know very well that *our* model has such and such shortcomings, a professionally developed model,

however, could solve the problem"); (b) let students experience genuinely beneficial consequences for subordinating mathematical knowledge to ethical reflections,
and (c) allow students to experience such subordination in sufficient frequencies.
This last point is particularly important because

> the subjective experience of those hundreds of hours [with mathematical knowledge as the
> sole warrantor of good consequences] may exceed the ideological parameters whilst
> remaining in the service of those ideologies by making us believe them through the sheer
> force of habitual action (Brown, forthcoming).

To provide the possibility that these reflective processes can effectively undermine the 'sheer force of habitual action', de Freitas (2008) suggests the development of a code of ethics of mathematical application that could inform
mathematical modelling processes:

> Why not construct an ethics of mathematical application, as we have for medicine? The
> application oath might simply demand that the mathematical agent (be it a student or a
> teacher or other) must reflect on the ethical consequences of her/his mathematical actions in
> the 'real' world, and seek to serve 'real' others in need of assistance through the use of these
> powerful tools [...]. [T]hen time spent in our classrooms on ethical reflection will serve the
> social justice goals of critical pedagogy (p. 92).

We argue that the first step in the realisation of the ambitious theoretical project
to develop an 'application oath' as an ethics of mathematical application is a further
clarification of the theoretical concepts at stake. This is why, for the time being, we
solely want to sensitise for the necessity of further theoretical considerations by
posing three simple, yet not explicitly discussed, questions: (1) *What qualifies a
reflection as an ethical reflection? Or with regard to our planed endeavour: What
qualifies a reflexive theory as an ethics?* (2) *What is the object an ethics is dealing
with? And* (3) *What is the relation between ethics and morality?*

In the next section, we approach these questions by re-contextualising selected
works by the sociologist Luhmann (1991) on the relationship between ethics and
morality for the field of mathematics education.

3 Towards an Ethics of Mathematical Application

What does it mean if we categorise an action, or a communication as bad or good *as
a whole* (as opposed to categorising a particular dimension of it); and what does it
mean to evaluate an action or communication as good or bad *as such?* What do we
mean if we say that one simply should not act or communicate in this and that way?
How is it possible to justify universal judgments like these, or are they even
justifiable at all? The specific forms of reflection that are indicated by these
questions lead into the sphere of what is commonly known as *ethics* (Tugendhat
1984). Any ethics is a "theoretical reflection of morality" (Luhmann 1991, p. 83).
That is, an ethics aims to reflect on the conditions of moral communication. This
means that any ethics stands in a theory-practice relationship to moral

communication and thus depends on the prevailing concepts of morality that are contingent upon the socio-historical conditions in which they are actualised.

If we understand ethics as a theoretical reflection of the empirical practice of moral communication and if, moreover, the social actualisations of morality are contingent, the first step towards an ethics of mathematical application is to provide *an empirical concept of morality*, which takes into account that the forms of moral communication are changing in time:

> I understand by morality a special form of communication which carries with it indications of approval and disapproval. It is not a question of good or achievements with respects, e.g. as an astronaut, musician, researcher or football player, but of the whole person insofar as s/he is esteemed as a participant of communication. Approval or disapproval is attributed typically to particular conditions. Morality is the useable totality of such conditions at any time. (Luhmann 1991, p. 84)

Firstly, this definition of moral communication does not refer to arbitrary entities but precisely addresses persons and persons only. Further, that is done in a very specific way: Everybody who communicates morally indicates (at least implicitly) the conditions under which he or she can or cannot approve *another person as a whole*. Moral communication thus always expands the range of its validity beyond what it initially sets out to evaluate. It unwittingly expands the approval or disapproval of a person's action, or communication to the person's entirety. Moralisation always entails generalisation. Conceptualised in this way, morality can be considered as the "conditions of the market of approval" (ibid.). Secondly, this empirical shift has the advantage that it limits moral communication to a very specific form of empirically observable communication. Therefore, we can ask

> what happens if conditioning of whatever kind (whether legal, political, racial or of personal taste) is moralized, with the consequence, for instance that X considers he cannot approve of Y and cannot invite him if he has a bust of Bismarck on his piano [or voted for Donald Trump to take a more recent example] (ibid.)

Moral communication is a very *specific* form of communication that, nevertheless, can be *universally* applied: Since we can (and often have to) ground our communications on distinctions different from the distinction between morally good and morally bad, not every communication is a moral communication. This indicates that moral communication is specific. However, when we conduct a moral communication, e.g. by using the moral code to communicate the conditions under which we approve another person, the behaviour which is subjected to moral evaluation can principally origin from all different kinds of realms. In other words, we are able to *morally re-code* all forms of behaviour or communication and evaluate them from a moral point of view. That is what equips the moral code with its universality.

Under the presumption that the moral code is universally applicable, it should be possible to apply it to itself. The question which emerges now is whether the distinction between good and bad is good or bad itself. Here, it is important to note that we *cannot* answer this question from the inside of morality because every binary code results in paradoxes in case that it is applied to itself. We simply cannot decide when we are using the distinction between good and bad to communicate morally if it is morally good or bad that we are doing so. In case that we try to approach this question *from the inside of the moral code*, the only answer we can get is a paradoxical one: The moral code as the distinction between good and bad is good if and only if it is bad.[3]

It seems that it is impossible to guarantee *from the inside of morality* that the application of the distinction between good and bad is itself good. This serves as an indication that there are social situations in which moral communication is rather counterproductive, e.g. when people rule out any contradiction by moralising. This intuition is supported by the acknowledgment of the finding that moral communication tends to provoke "over-engagement of the participants" (Luhmann 1991, p. 86) and thus "is close to conflict" (ibid.) or even violence:

> Whoever communicates morally by making known the conditions under which he disapproves of others and of himself, invests and places at risk his self-approval. (ibid.)

Therefore, an ethics also has the task to define, and thereby limit, the space of applications of any moral communication. This means that an ethics should explicate the conditions under which it is good to use the corresponding moral code to evaluate particular communications and, which maybe is even more important, the conditions under which it is not good to do so. We can exemplify the limitations of moral communication by shortly describing what it would mean for the practice of mathematics as a scientific discipline if it were overdetermined by the moral code. Mathematical communication is organised in the *general medium of proof*. That is to say, every particular theorem which is presented inside the community of mathematicians is always accompanied by a proof. The theorem is then evaluated by means of the distinction between true and untrue due to the validity of the proof. Even if we only presuppose a very weak conception of truth—such as truth is what is counted as true by the community of mathematicians—we immediately see how fatal it would be to identify true mathematical theorems with the side of 'the morally good' and untrue mathematical theorems with the side of 'the morally bad'. It would be fatal because any progress in mathematics is based on the interplay between both *proofs and refutations* (Lakatos 1976). Therefore, we simply cannot brand a mathematician who presents a false theorem as morally bad because this would paralyse the research practice as a whole. This does not mean that mathematical communication is not subject to certain moral conditions that can be articulated in an ethics (c.f. Hersh 1990), but it only means that the practice of mathematics as a scientific discipline simply cannot "be integrated into the social system by means of morality" (Luhmann 1991, p. 85). This implies that scientific mathematical communication is organised by the functional code true/untrue and

---

[3]Gödel (1931) used this insight and developed a method by means of which it is possible to construct undecidable propositions in any sufficiently rich formal system, e.g. the arithmetic of natural numbers (Incompleteness theorem I).

this code operates at a higher level of *amorality*.[4] Here, amorality does not signify the opposite of "good", but the negation of the distinction good/bad itself. So, in order to be productively applicable, the functional code true/untrue must necessarily remain at a certain distance to the moral code good/bad. Again, that does not mean that moral communication on mathematics is impossible, but it does only mean that the moral code would be highly dysfunctional. It would be dysfunctional as the comprising code of mathematical communication since it would simply undermine the practice of mathematics: Proofs as the exchange medium of mathematical knowledge "emerge through the process of proposal and criticism through which they are improved enough to withstand the critical attitude of mind" (Ernest 2001, p. 278).

So far, what is our interim conclusion with respect to our aim to develop an ethics of mathematical application? Firstly, mathematics is neither intrinsically good, nor intrinsically bad. Therefore, it is important to condemn any ethics that is promoting an all too easy solidarity of mathematics with one side of the moral code. Secondly, an ethics of mathematical application must thematise morality *as a distinction*; that is, moral communication is a form of communication that re-codes decisions in a mathematical modelling process based on an attribution of the label of the morally good *or* bad. Thirdly, we must acknowledge the very specific nature of the moral code as being universal and specific at the same time. Consequently, this means that an ethics of mathematical application should also *reflect upon the limits of the scope of moral communication on mathematical application*.

At this point, we reach a level of abstraction that brings with it specific theoretical challenges because we must rely on a theory that is able to distinguish between the use of different distinctions. In our case, this means that we have to find ways *to agree upon acceptance and rejection values in relation to our moral distinction between good and bad* (Luhmann 1991, p. 85). In other words, we have to find a way to negotiate a consensus of when it is productive to moralise, and when it is unproductive to do so. But how could these acceptance and rejection values look like? And how can we agree about these conditions?

Here, we follow de Freitas (2008), who argued that we always have to consider the possible consequences of mathematical actions for the 'real' others in the 'real' world when we want to evaluate our decision making in the modelling process from a moral point of view. Thus, the ethical task is to transcend our individual position as decision makers in the modelling process (including our individual or commercial interests) and observe the consequences of our models from the perspective of others who could be affected by the decisions within the processes of modelling

---

[4]This argument has to be generalised as we are living in functionally differentiated societies where *all* functional systems "owe their autonomy to their individual functions, but also to their binary codes" (Luhmann 1991, p. 85), while in "neither case can the two values of these codes be made congruent with the two values of the code of morality: In case that we, for example, consider the distinction between government and opposition in democratic political systems", we "do not want the government to be declared structurally good and the opposition structurally bad or evil" (ibid., p. 85f).

that were realised in practice. However, these reflections should not simply accompany the modelling process, but instead they should already inform or mediate our practical decisions in the construction of the mathematical model. Although it is incontestable whether the shifting of perspectives is a central theoretical figure to all moral considerations, we should not forget that the reflection of consequences of a mathematical model at the stage of its construction is not a straight-forward endeavour because it demands a very specific form of reasoning. The reasoning that is required can be characterised by the following form: If p then q, although p is not the case. Skovsmose (2008) calls this form of reflection "hypothetical reasoning" (p. 165) because it (re-)inscribes the practical consequences of a theoretical action into the theory itself, but not into practice. It remains theoretical and is thus only hypothetical. In this way, it is possible to reflect upon the gap between theory and practice in theory. However, the immanent limitations of such a theoretical approach, which aims to include the practical consequences of theoretical considerations *in theory* are clearly articulated by Habermas (1973) in his famous book *Theory and Practice*:

> Of course, the objective application of a reflexive theory under the conditions of strategic action is not illegitimate in every respect. It can serve to interpret hypothetically the constellations of the struggle [...]. Seen from that anticipated goal, such interpretations are retrospective. Therefore, for strategic action and for the maxims by which the decisions in the discourse that prepares for this action are justified, these interpretations open up a perspective. But the objectivating interpretations themselves cannot claim a justificatory function; for they must comprehend counterfactually one's own action, which now is only being planned [...]. (p. 40)

In other words, the first step towards hypothetical reasoning requires that we have to anticipate future consequences. In a second step, this projection allows us to retroactively evaluate our decisions from this *fictive point* in the future, although all decisions are yet to be made in reality (or will possibly never be made). Therefore, "the objectivating interpretations" are bound to a projection of hypothetical consequences into the future and thus, due to their hypothetical nature, "cannot claim a justifying function". Given the case that we cannot justify our actions in this way, on which criteria can we base our decisions then? Since we can only retroactively reflect upon the definite practical consequences of certain theoretical decisions in the modelling process, that is, *after* the decisions as well as the model are put into practice, we are thrown back to the question of the motivation of moralised decisions.

Although we have shown that we cannot sufficiently justify our decisions by hypothetical reasoning, we can still argue that it might be possible to initiate a democratic negotiation process to solve the justification problem—at least with regard to *the moral intention of a decision*. In other words, we could negotiate whether the intention of a decision in a modelling process is morally good or bad. Then, we would commonly decide whether we confirm the decision or not. However, this approach becomes invalid as soon as we take into account that good intentions can have bad consequences and vice versa. Therefore, Luhmann (1991) asks:

> If reprehensible action can have good consequences, as the 17th and 18th century econ-
> omists assure us, and if inversely the best intentions can lead to bad results, as we can see
> from politics, then moral motivation blocks itself. Should ethics then counsel good or bad
> action? (p. 87).

The problem of justification already manifests itself in the motivation of our morally re-coded actions *because the depicted mode of reflection does not provide any criteria that could sufficiently inform our decisions*. Therefore, we pose two questions: (1) Is an ethics that guides moral communication on mathematical application possible at all? And (2) How could it regulate and deregulate moral communication?

To come straight to the point, we do not know the answer to these questions. What we believe is that any serious attempt to develop a positive conception of an ethics of mathematical application has to satisfy certain theoretical minimal conditions that we have, at least rudimentarily, explored in this section. Thereby, we approached the question *ex negativo* in order to show how an ethics of mathematical application could look like. That is, we focused on the identification of selected theoretical dead-ends to illustrate how an ethics of mathematical application *cannot look like* rather than to give a positive draft of it:

1. Any attempt to identify mathematics and its applications with just one side of the moral code can only be condemned as a 'quasi-ethics' because the moral code has to be addressed by an ethics of mathematical application as distinction with two sides.
2. Since it is impossible to justify the application of the moral code from the inside of morality, the limits of moral communication to evaluate processes of mathematical communication and action have to be articulated by the identification of rejection and acceptance values of moral communication on mathematical application (This is what led us to the problem of justification).
3. Neither the consideration of hypothetical practical consequences ('hypothetical reasoning'), nor the reflection on the moral intention could provide us with unambiguous decision-making guidelines that could sufficiently justify theoretical decisions in the mathematical modelling process.

However, we still believe that there simply is no alternative to a reflexive approach to an ethics of mathematical application. That is, an ethics of mathematical application has to remain a *theoretical reflection of the moral communication on mathematical application*. In the next section, we approach the conceptual problems from a more practical point of view by asking the following questions: How is moral communication on mathematical application structured in contemporary late-modern societies outside the realm of schooling? Moreover, which ethics (if any) informs the moral communication on mathematical application in these social spheres?

4   The Ethics of Solutionism

In apparent contrast to our developed thesis that mathematisation is always accompanied by 'ethical filtration', Morozov (2013a) argues that the promotion of the formatting of our social life by means of mathematisations often does not happen in ignorance of moral considerations, but exactly in the ethos of making the world a better, safer, greener, and more equitable place. For example, the former CEO of Google, Eric Schmidt, frames his inspiration: "Technology is not anymore about hardware or software. It is about collecting and analysing enormous masses of data in order to change the world to the better" (cited in Morozov 2013a, p. 9, translated by H.S-P.); Mark Zuckerberg, the founder of Facebook, strikes a similar tone: "We do not wake up in the morning to earn money". Rather, Facebook follows the mission of "making the world more open and interconnected" (cited in Morozov 2013a, p. 9, translated by H.S-P.). Here, it is important to note that the leaders of digitalisation do not position mathematisation as *one* form of regulating social practices amongst others. They do not treat it as a specifically motivated form that is contingent upon a particular distinction which only takes into account those characteristics of a practice that can be quantified successfully. Instead, they paint the mathematisation of the social in the colours of a completely unideological endeavour:

> Out with every theory of human behavior, from linguistics to sociology. Forget taxonomy, ontology, and psychology. Who knows why people do what they do? The point is they do it, and we can track and measure it with unprecedented fidelity. With enough data, the numbers speak for themselves (Chris Anderson, editor in chief of the Wired magazine 2008, cited in Han 2014, p. 99).

In other words, Anderson discredits *any* theory of human behaviour as inadequate to explain human behaviour. He thus aligns the distinction "numerical determination of human behavior"/"(theoretical) explanation for human behavior" to the distinction adequate/inadequate. As Morozov's analysis makes us aware, the distinction adequate/inadequate is, however, already morally coded within the ideology of solutionism. Any theoretically informed explanation of human behavior is thus morally discredited. The 'objective' analysis made possible by 'Big Data' is, in turn, posited as the *unideological opponent*. However, as Žižek (1989) frequently reminds us, the position that declares itself completely free of ideology is the one that we should be most suspicious of because "the idea of the possible end of ideology is an ideological idea par excellence" (p. xxiv).

Any semantics that reifies its own way to describe society within the society to the only one can be labelled as an ideology. Consequently, Morozov (2013b) calls this form of reasoning, which is "[r]ecasting all complex situations [...] as neatly defined problems with definite, computable solutions [...] if only the right algorithms are in place" (p. 5), an 'ideology of solutionism'. *Firstly*, the semantics is labeled by Mozorov as 'solutionism' because social problems are conceived as sort of puzzles that are, in principle, solvable only if enough data is collected and analysed; *secondly*, the semantics is indicated as an 'ideology' because it disavows its own "political foundation" (Žižek 2000, p. 169); *it reifies one way of seeing the world to the only one.*

At first glance, it could be assumed that the semantics of mathematisation is again grounded in an 'absolutist' conception of mathematics, so that its presupposed universal applicability guarantees the superiority of mathematisation. The sociologists Espeland and Stevens (2008), who developed a sociology of quantification, argue in alignment with this idea when they state that "quantification facilitates a peculiarly modern ontology in which the real easily becomes coextensive with what is measurable" (p. 432). This supposed 'modern ontology' identifies *being* with mathematisation and *nonbeing* with non-mathematisation and thus supports the mathematisation of the social. However, the focus on an ontological conception of mathematics disavows that the mathematisation of the social is embedded in a very specific moral horizon. This moral horizon suggests that any attempt to counteract the solving of problems by means of mathematisation *is a reactionary intervention against the human(e) project of making the world a better place.* In other words, the backside of mathematisation (that is: non-mathematisation), which is standing for the possibility of an extra-mathematical answer to a social problem, is blanked out by its alignment to the backside of the moral code. Consequently, from the inside of the ideology of solutionism, anybody who criticises the *self-referential closure of the social process of mathematisation*[5] is immediately stigmatised as morally bad. Therefore, the moral horizon as well as its self-expression as unideological is a necessary support for the effectiveness of this ideology as it is only by the horizon of the potential realisation of a better world that it can justify that humanity should let a group of supposed pioneers solve its problems, even before these problems have been identified *by human experience,* and before they have been *problematised* in the political arena.

Going one step beyond Espeland and Stevens (2008), we thus argue that we cannot understand the semantics of mathematisation without the consideration of the moralisation it elicits. This moralisation becomes necessary precisely because the classical ontological distinction between *being* and *nonbeing* fails as an all encompassing semantical figure. It appears that it is not possible to simply identify the side of being with the entities that are quantifiable, and the side of non-being with the entities that are not quantifiable. This very failure of the modern ontology as an all encompassing semantical figure effectuated a moral supplementation of the ideology, that is, the linkage of mathematisation with the side of the morally good. However, we have already extensively argued in the section above that an ethics of mathematical application has to thematise the moral code with respect to mathematical application *as a specific distinction* instead of universally identifying it with the side of the morally good (*or* the morally bad). Thus, the ethics that supports the

---

[5]This argument of a self-referential closure of the process of mathematisation refers to its circularity: As soon as a mathematisation is implemented in a social practice, it can only be substituted by another, and possibly more sophisticated, mathematisation.

one-sided moralisation of mathematisation is in fact another example of what we have earlier called a 'quasi-ethics'.

If mathematical application in school mathematics, and the mathematisation of the social are both in many cases informed by 'quasi-ethics', it becomes necessary to ask: What are the similarities and differences between the two identified quasi-ethics? Moreover, what consequences can be drawn for our project of an ethics of mathematical application?

5 Concluding Remarks

The semantics we have unveiled have tried to establish mathematical application as morally good. This accounts both for the 'quasi-ethics' that informs mathematical application in school mathematics on the one hand, and the 'quasi-ethics' supporting mathematical application in the sphere of the social beyond schooling on the other hand. However, they do so in different ways: In school mathematics, mathematical knowledge is established as morally good by making the application of mathematics have beneficial consequences for the students. Such identification of mathematical application and the morally good tends to constitute an 'absolutist' conception of mathematics because 'true', universally applicable mathematics remains as *always out of reach* (cf. Sect. 2). In the sphere of the social, the superiority of mathematisation that is applied to solve social problems *is justified by moralisation*. Any non-mathematical approach to the regulation of the social is marked as morally inferior, so that anybody who is not willing to participate in the mathematisation of the social not only rejects a particular way of seeing the world, but seemingly refuses to participate in the global struggle to make the world a better place for everyone. Within this line of argumentation, we immediately rediscover the two attempts to justify moral judgments that we explored above. The 'quasi-ethics' of school mathematics retroactively establishes mathematical application as morally good *by means of its practical consequences*, while the quasi-ethics of mathematisation does so *by means of its moral motivation or intention* (and thereby excludes those who are not willing to share these intentions). School mathematics and solutionism thus differ in the terms by which they establish a 'quasi-ethics' of mathematical application.

Nevertheless, what the two depicted 'quasi-ethics' have in common is that they promote an alignment of mathematical application and the morally good. Further, in doing so, they also share a common blind spot: The 'quasi-ethics' become blind to the primordial distinction that opens up the whole field in which they operate. In both cases, this is the distinction between mathematisation and non-mathematisation. In other words, both ethics are 'quasi-ethics' because they are blind to the possibility of a non-mathematical approach to the social. Therefore, we can identify *the distinction between mathematisation and non-mathematisation as the extra-mathematical, political foundation of any application of mathematics.*

This allows us to complement our suggestions towards the implementation of moral communication on mathematical application in the teaching and learning of mathematics. It is not only important that the students are not credited when they relativise their constructed models (cf. Sect. 2), but it might be even more important to "strongly reject any conceit, scientific or otherwise, that measurement provides privileged or exclusive access to the real." (Espeland and Stevens 2008 p. 432). This means that the primordial decision, *whether or not* the application of mathematics is a useful way to treat a certain problem, cannot be predetermined in advance and thus be deprived from the responsibility of the students. The distinction between mathematisation and non-mathematisation has to remain *inside the scope of reflection during the entire modelling process in mathematics education.* This means that a reasonable decision *against* a mathematisation of the problem should not be excluded from the space of 'positively credited' communications about a modelling problem. However, this immediately leads to the following question: How is it possible to evaluate a decision against a mathematical approach to a problem as reasonable or non-reasonable?

The only possible answer here is to rely on distinctions that differ to the one between mathematisation and non-mathematisation. For example, the distinction between morally good and morally bad. In other words, the moralisation of the problem at stake is *one* possibility to identify the cases in which it is productive to subordinate the problem to a mathematisation and the cases under which it might not be productive to do so. In the example of Skovsmose's Micro-Society project, students could be promoted when they refute a mathematisation of social welfare benefits, e.g. by arguing for an unconditional basic income as something that is morally good (i.e., a decision based on a moralisation of the situation that can of course be questioned from a deviating moral horizon).[6]

As we have shown above, to identify something as morally good is never a straightforward endeavour and must be (re-)evaluated case-by-case. Moreover, moral communication cannot be an end in itself, that is, it is impossible to argue for moralisation as the ultimate ground of all decisions with regard to the mathematisation of the social. The moral code itself is also only *one* distinction amongst others which means that we have only shifted the problem of justification. The moral code is a distinction that can inform the primordial distinction between mathematisation and non–mathematisation that is inscribed into any particular application of mathematics. At the same time, it is *unable to limit its own scope of application.* It is precisely here that we enter the field of ethics as a theoretical reflection of the conditions of moral communication on mathematical application. Furthermore, it was one of the identified key challenges to an ethics of

---

[6]Rejecting mathematisation *as such* is an identification of non-mathematisation as good (and mathematisation as bad) which is structurally completely equivalent to solutionism. However, in a semantic environment in which mathematisation and quantification have become the one and only legitimate sources for moral judgment, even rejecting mathematisation *as such* becomes a political act At least it yields the possibility to argue outside mathematisation (a similar argument would account for a semantic environment governed by "anti-mathematisation").

mathematical application that it has to *warn against moralisation and thus provide orientation to decide under which conditions it is productive to evaluate mathematical applications from a moral standpoint and under which conditions it is not.*

This warning function, we claim, can be particularly productive for those approaches within mathematics education that seek to re-politicise school mathematics, e.g. critical mathematics education. An ethics of mathematical application could be employed as a means for reflexively controlling the necessary moralisation that inevitably comes along with a re-politicisation. In this way, an ethics could serve as a 'reflective warning system' that prevents undesired coalitions with solutionism and thus helps to recover "the meaning of 'critique' in critical mathematics education" (so the title of: Pais et al. 2012).

An ethics developed in this spirit should by no means be confused with a mere relativism in the sense that we simply cannot make any decisions at all but sensitise for the hypothesis that any attempt to identify *rejection and acceptance values of moral communication on mathematical application* can only be grounded in the social conditions of late-modernity itself which need to be investigated from a semantical *as well as* structural perspective. In this chapter, we tried to exemplarily reconstruct the semantics that inform about moral communications on mathematical application in the field of school mathematics and the field of the social. Further, we presented a theoretically, yet rather naïve, analysis of the ideology of solutionism. Recent developments of ideology critique in the field of mathematics education (e.g. Pais 2017; Lundin and Storck Christensen 2017; Straehler-Pohl 2017) appear to provide a profound analytical frame for this purpose. So far, we paid almost no attention to the very specific social structure of late-modern societies that must be considered as well in order to systematically develop an ethics of mathematical application. Such an analysis could inform about moral communication on mathematical application in (critical) mathematics education.

Therefore, we conclude in suggesting firstly, developing an ideology critique of the semantics of solutionism as a desideratum; and secondly, a further development of our outlined framework of an ethics of mathematical application that takes into account both the semantics and social structure of the contemporary late-modern society.

References

Brown, T. (forthcoming). Concepts and commodities in mathematical learning. In A. Coles, E. de Freitas, & N. Sinclair (Eds.) *What is a mathematical concept?* Cambridge, UK: Cambridge University Press.

Espeland, W. N., & Stevens, M. L. (2008). A sociology of quantification. *European Journal of Sociology, 49*(03), 401–436.

Ernest, P. (2001). Critical mathematics education. In P. Gates (Ed.), *Issues in mathematics teaching.* New York: Routledge.

De Freitas, L. (2008). Critical mathematics education: Recognizing the ethical dimension of problem solving. *International Electronic Journal of Mathematics Education, 3*(2), 79–95.

Gödel, K. (1931). Über formal unentscheidbare Sätze der Principia Mathematica und verwandter Systeme I. *Monatshefte für Mathematik und Physik, 38*(1), 173–198.

Han, B.-C. (2014). *Im Schwarm. Ansichten des Digitalen.* Berlin: Matthes & Seitz.

Han, B.-C. (2017). *In the swarm: Digital prospects.* Cambridge, MA: MIT Press.

Hersh, R. (1990). Mathematics and ethics. *Humanistic Mathematics Network Journal, 1*(5), 9.

Habermas, J. (1973). *Theory and practice.* Boston: Beacon Press.

Lakatos, I. (1976). *Proofs and refutations: The logic of mathematical discovery.* Cambridge: Cambridge University Press.

Luhmann, N. (1991). Paradigm lost: On the ethical reflection of morality. *Thesis Eleven, 29*(1), 82–94.

Lundin, S. (2012). Hating school, loving mathematics: On the ideological function of critique and reform in mathematics education. *Educational Studies in Mathematics, 80*(2), 73–85.

Lundin, S., & Storck Christensen, D. (2017). Mathematics education as praying wheel: How adults avoid mathematics by pushing it onto children. In H. Straehler-Pohl, N. Bohlmann, & A. Pais (Eds.), *The disorder of mathematics education. Challenging the socio-political dimensions of research* (pp. 19–34). Cham: Springer.

Morozov, E. (2013a). *Smarte neue Welt. Digitale Technik und die Freiheit des Menschen.* München: Karl Blessing.

Morozov, E. (2013b). *To save everything, click here: The folly of technological solutionism.* Public Affairs.

Pais, A. (2017). The narcissism of mathematics education. In H. Straehler-Pohl, N. Bohlmann & A. Pais (Eds.), *The disorder of mathematics education. Challenging the socio-political dimensions of research* (pp. 53–63). Cham: Springer.

Pais, A., Fernandes, E., Matos, J. F., & Alves, A. S. (2012). Recovering the meaning of "critique" in critical mathematics education. *For the Learning of Mathematics, 32*(1), 28–33.

Skovsmose, O. (1994). *Towards a philosophy of critical mathematics education.* Dordrecht: Kluwer.

Skovsmose, O. (2008). Mathematics education in a knowledge market: Developing functional and critical competencies. In E. de Freitas & K. Nolan (Eds.), *Opening the research text: Insights and in(ter)ventions into mathematics education* (pp. 159–188). New York: Springer.

Skovsmose, O., & Borba, M. (2004). Research methodology and critical mathematics education. In P. Valero & R. Zevenbergen (Eds.), *Researching the socio-political dimensions of mathematics education: Issues of power in theory and methodology* (pp. 207–226). New York: Springer Publisher.

Straehler-Pohl, H. (2017). De|mathematisation and ideology in times of capitalism. Recovering critical distance. In H. Straehler-Pohl, N. Bohlmann, & A. Pais (Eds.), *The disorder of mathematics education. Challenging the socio-political dimensions of research* (pp. 35–52). Cham: Springer.

Tugendhat, E. (1984). *Probleme der Ethik.* Stuttgart: Reclam.

Žižek, S. (1989). *The sublime object of ideology.* London: Verso.

Žižek, S. (2000). *The ticklish subject: The absent centre of political ontology.* London: Verso.

How to Be a Political Social Change Mathematics Education Activist

Peter Appelbaum

Abstract Three sets of nomadic epistemological categories (Deleuze and Guattari) that coexist with other theoretical frameworks of mathematics education discourse and practice are used to suggest an approach to changing ourselves as mathematics educators through the ways that we think and act. The argument is that these reconceptualizing processes can change our worlds of possibility for mathematics education while allowing coexistence with more mainstream programs of research and practice: Arendt's work, labor and action; Pitt's youth leadership, voice and participation; and McElheny's architectural, scientific, and artistic models. Such epistemological categories establish topologies, reconstructing subjectivities in the process—a tactic of alterglobal social movements that potentially politicize mathematics education: we change ourselves to change the world. Psychoanalytic responses to the terror of change, and the need to address the legacies of mathematics as a component of colonialism, are considered as components of broader social change.

Keywords Politics of mathematics education · Nomadic epistemology
Social justice · Alterglobal education

1 Introduction

My title is at once a fantasy of what might be accomplished by those who would join me in this conceptual game of rethinking where and how mathematics education can take place, become, and change, and also a serious invitation to try out a mostly philosophical strategy of adopting non-mainstream language. Together, we might help mathematics education more explicitly address political and social issues. I propose that striving to change ourselves as mathematics educators through consciously using new epistemologies remakes our realities. My proposition is that

P. Appelbaum (✉)
Arcadia University, Glenside, PA, USA
e-mail: appelbap@arcadia.edu

© Springer International Publishing AG 2018   53
M. Jurdak and R. Vithal (eds.), *Sociopolitical Dimensions of Mathematics Education*, ICME-13 Monographs, https://doi.org/10.1007/978-3-319-72610-6_4

remaking ourselves in this fashion bypasses a crisis of mathematics education, and education in general, which is grounded in the fantasy of reforming practice toward a dream of power, a dream of putting into place a near-perfect curriculum, if not at least a well-planned set of steps in what would be called 'the right direction'. Instead of planning what to implement, we change how we see and think and 'be' as mathematics educators. This process does not make a utopia, a perfect world that cannot exist, but rather co-exists with that always-imperfect and ever-changing fluidity we sometimes call 'life'.

One can read this chapter as a tongue-in-cheek manual for how to coexist with typical school mathematics learning contexts, still striving to change ourselves as mathematics educators by consciously using new epistemologies to remake our realities, and supporting the efforts of those in these environments, while simultaneously working toward social change. This format is a non-traditional approach to sharing the synthesis of a number of action research projects that have studied the ways that mathematics education communities, mostly outside of traditional schools, are bubbling forth, taking action, and supporting social justice movements. Such study can often help us to re-enter school mathematics education in ways that highlight social and political dimensions. These research projects used experiences outside of formal schooling to generate concepts that were later brought into the planning of school programs, the design and evaluation of these programs, and the forms of instructional organization in these programs. At first, Community arts groups (Spiral Q Puppet Theater "SparQ" groups, Philadelphia, USA), university courses (Undergraduate seminars in the US that are not situated in mathematics or in education), global online hacker communities and other social media spaces, do not readily appear to be sites of "mathematics education". Nevertheless, when we think about these locations through a mathematics education lens, we can identify forms of teaching a research, reflecting upon or inventing mathematics, and/or mathematics learning strategies. Following such retrospective redefinition of what did not appear to be mathematics teaching and learning as varieties of authentic mathematics learning experiences, the next step was to then look at school mathematics environments with the lens of the out-of-school characteristics. In this way, political and social potential of teaching, research, and mathematical activities taking place in more traditional, school environments were more accessible to analysis and interpretation. This reflexive, community-based, participatory research (Leavy 2017) generated the theoretical framework and potentially transferable conception of collaboration presented in the following sections.

Three examples are used in this chapter to illustrate how 'nomadic epistemologies' (Deleuze and Guattari 1987) construct new topologies of time and space; the new topologies can work in tandem with 'alterglobal social movements'. Alterglobal movements (Appelbaum and Gerofsky 2013; Pleyers 2010) theorize communities in flux that coalesce and take action without requiring fixed identities, clear goals, justification, or defined structures, while maintaining a strong commitment to ethical principles of inclusion, diversity, human recognition and dignity (Butler 1997, 2010). The latter phrase, referring to recognition and dignity, may feel like familiar goals. However, in the context of globalization, it has become necessary to maintain a focus on these goals, as it becomes critical to redefine the meaning of such terms and the means toward them, in the light of transnational business interests, global information flow, mass migration and refugee movements, and new uses of technology for political transformation. The term 'alterglobal' has come to represent various forms of collective action responding to the negative aspects of globalization—corporate personhood, the dominance of markets over ethics, the increasing need to understand diaspora identities, the changing nature of identity through social media, and so on—in ways not naively 'anti-globalization' (which would treat globalization as something to work against). It is also distinct from 'counter-globalization' (working in opposition to globalization), 'super-globalization' (overcoming globalized interactions), or 'hyper-globalization' (taking globalized interactions to an extreme). Alterglobalization calls for renewal of political citizenship and activism, bypassing traditional ideas about how to make social change. For example, alterglobalization may reject traditional ideas of creating revolution, whether by peaceful or violent means, which usually assume that people live within nation states, or that change happens within the confines of national boundaries. Social movements such as the Green movement, Feminist movements, Slow Food and Slow Clothing movements, and Animal Rights actions, all have in common the ways that they transcend countries and work inside and outside of the marketplace.

The fundamental questions of mathematics education might coalesce for this moment around dispossession and recognition. Dispossession describes the condition of those who have lost land, citizenship, property, and a broader belonging in the world. Recognition refers to the visible status of a human being participating in a collection of communities. How can mathematics education work with, within, and in parallel to crowds of recognition, to protest dispossession through performative, trans-national projects? How can we use commitment to alterglobal action in supporting complicated conversations that are recognized and that recognize? How can we nurture maker-communities of conviviality, designing, and committed aesthetics? How does mathematics education strengthen recognition and when does it weaken dispossession?

2 Three Discourses

My strategy has four components: (a) Use language both inside and outside of the common discourse; (b) make clear to ourselves and others that these new languages do not confront existing pedagogical terminologies, but rather live alongside them; (c) carry out research and practice grounded in the new languages, thus transforming the focus and meaning for the research and practice, still maintaining support for current practices; the power of the new discourses changes how we understand ourselves and the worlds of mathematics education in which we take part; (d) finally, describe for ourselves and others how the new research projects and practices have shifted and altered relations of power, privilege, recognition, dignity, authority, knowledge, and mathematics (Appelbaum 2017). The next three sections of this chapter describe discourses that can be used to change ourselves, and hence our worlds. The first comes from the philosopher Hannah Arendt.

2.1 Hannah Arendt: Work, Labor, Action

Arendt (1958) described activity as taking on the character of work when the activity is intimately related to conditions of life. The same activity would be called "labor" if the effort and time were removed from those conditions of life and of being human, or, in contrast, "action", if people come together in community through the activity. For example, if one makes a chair in order to sit on it, the making of the chair is work. If one makes a chair in a factory so that the corporation can sell the chair to others, then the making of the chair is labor. If groups of people make chairs so that they can sit together and talk, eat, sing, and so on, then the making of the chairs is action.

We can use this language to describe the activity in a mathematics classroom, or other learning context. How is the use of mathematics related to the relationships in this collection of people? For example, if a learner is practicing the application of an algorithm, then this learner is working. If the work is only serving others, such as when learners produce materials for assessment, then this is labor, and not genuine "work". If the work contributes to community building, then the activity is action. Arendt's categories of analysis would allow conversations about classroom community to connect with issues of power and knowledge in a broader social context, alongside more commonplace analysis of performance related to content objectives, the development of mathematical thinking, forms of argument, and other empirical research. In this way, the analysis of mathematics classrooms and the forms of knowledge production would be able to contribute to the ongoing development of pedagogical methods and the design of classroom materials.

Analyzing mathematics teaching and learning in terms of work, labor and action might seem like an entirely new way to study, assess or design mathematics curriculum. However, there is a strand of mathematics education research that has used analytic categories in similar ways. For example, Kirshner (2002) successfully analyzed the implicit metaphors of practices in terms of how they align with psychological learning theories. Each theory might align with contemporary professional recommendations, yet Kirshner noted that learning progresses in fundamentally different ways in these commonly held views: Habituated learning develops *incrementally*, conceptual construction involves *transformation* of existing conceptual structures from perturbations that arise out of reflective abstraction, and enculturation features *discontinuity* between prior patterns of participation and new cultural patterns appropriated through cultural enmeshment. Kirshner hoped for a richer understanding of reform in mathematics education practice through the use of his "cross-disciplinary approach" that centered these distinctions. I am proposing something akin to Kirshner's approach. Arendt's categories highlight the activity in the classroom as constituting power relationships connected to broader sociological and political structures in society. How the mathematics is produced, and toward what purposes in terms of the community, is something that stands apart from more typical mathematics education research and reflection. Analysis using these categories of work, labor and action might dialogue with more commonly employed forms of research and practice.

2.2 Alice Pitt: Youth Leadership, Voice, and Participation

Another set of analytic categories has been suggested by Pitt (2003), who described organizing the curriculum with the explicit goal of supporting each learner (a) to be in positions of leadership, (b) to be heard by others, and (c) to recognize that their participation in the collective makes a difference. Mathematics educators can design research and practice to emphasize these as well. These three foci offer alternatives to more commonly applied characteristics of curriculum, such as accountability, test scores, international comparisons –even content concepts and skill objectives. Ongoing assessment would be constructed to support the continued reflection on who is and is not having access to leadership, who is or is not recognized and heard in the group, and how and when various learners are participating in the group activities. Lessons and activities would be designed to support the nurture of student contributions of ideas during the activity, so that their ideas transform what follows in a very fundamental sense. Sharing of conjectures, questions, modifications of problems, applications of mathematics, and so on, would necessarily be the result of students' sincere and committed demands on the group. Leadership beyond the classroom, in the broader community, would be considered essential as a component of the ongoing work.

This second set of alternative or simultaneous categories of analysis might work in tandem with the categories from Arendt, or independent of them, just as they might be parallel to or in dialogue with more common foci of mathematics education research. They address what Mellin-Olsen (1987) wrote about in his introduction to his seminal work, *The Politics of Mathematics Education*, as the context-level of behavior in the learning theory, that is, the possibility of explaining how behavior of students can be interpreted in terms of the types of learning situation. Mellin-Olsen noted that a learning theory lacking such a context-level can leave theory and associated practice searching for ways to "cure" the student. When this takes place, Mellin-Olsen concluded that the student has not been able to take advantage of the school as an opportunity. Instead, the type of learning situation has contributed directly to the potential (harmful?) "intensity of the school's influence" (p. 9). Pitt's categories extend Mellin-Olsen's work, indicating that there are parallel worlds of mathematics curriculum interacting with each other. One is traditionally studied, consisting of particular mathematical content, specific attention to mathematical forms of questioning, arguing, and convincing, and so on. At the same time, we have the world of work, labor and action discussed in Sect. 2.1 above.

We can further study the implicit curriculum of leadership, both social and cognitive, skills of participation in group process, and methods of sharing ideas and influencing others. How the student understands their own position relative to the group, and how the content of the curriculum and instructional methods of the teacher combines with this understanding to enact possibilities for action in this context, can be understood on its own terms. This opens up new directions for research and practice. Or, we can enact more complex forms of research and analysis, exploring how these mathematics education worlds are in relation to other types of data that might be collected about a student, such as test performance, choice of further study, and so on. While there is a large body of research demonstrating that human relationships can influence mathematics learning —family, school, classroom, community, and so on (Perry and Dockett 2002)—what these categories from Arendt and Pitt suggest is that there is a parallel set of ways that relationships specific to a mathematics education context are political. These other forms of human relation carry with them informal social and political 'lessons' about ways of behaving with others in association with knowledge and learning.

2.3   Josiah McElheny: Three Ways to Use Models

While the first two examples so far have seemed to direct mathematics education away from mathematics content per se, my next suggested set of categories provides a way to analyze the materials and images that we use to create representations of mathematical concepts. The sculptor McElheny (2007) described different stereotypical ways that various professionals tend to use models in their work. Mathematics teachers and students often work with models, whether they are visual representations such as circle diagrams for fractions, physical objects, base-ten blocks for place value, or flow charts for complex algorithms. Given this association, we might learn from McElheny, who notes that an architect can use models in any of these ways, but often specifically uses a (scale) model to convince others of the strength of their proposed design. In McElheny's reflection, a scientist mostly uses a model to analyze the relationships among the components of the model, and then to test hypotheses about these relationships. An artist, McElheny suggests, might also use any of these approaches; he notes however that an artist might be compelled to use a model to challenge our assumptions or to raise questions rather than answer them. In a classroom, teachers and students might similarly use models (a) to convince others of a conclusion they have come to, (b) to learn about new concepts, or (c) to raise new questions about the mathematics (Appelbaum 2012). Members of social circles, NGOs organizing projects, or political actors working to change public policy might similarly use mathematical and other models in various ways. Whether in a traditional classroom, or elsewhere, this set of characteristics can be used to create a variety of experience, to clarify the purposes of one's situation, or to address specific needs of the group.

The strategy of multiple representations has a long history in mathematics education, and continues to be promoted in many ways, from recent applications of

research to classroom practice (Tripathi 2008), to the power of multiple representations for inclusive practices targeting special learners (Sliva Spitzer 2003). Less
attention has traditionally been paid within mathematics education curricula to the
ways that mathematical representations implicitly format the sense of possible
actions that can be taken using the mathematics, or outside of the mathematics
being learned (Skovsmose 1994, Star 1995, Appelbaum 2012). McElheny's
categories promise the possibility of understanding multiple representations
simultaneously as cognitive tools for conceptual development, and also as
content-specific devices that carry with them the potential to be combined with the
political categories from Arendt and Pitt.

3 Changing Ourselves to Change the World: Making Our Work Explicitly Political

In each of the above examples, new sets of categories function as nomadic epistemologies. They could be used in traditional school mathematics contexts, or to
construct entirely new experiences in any social situation, whether face-to-face in a
neighborhood, or through social media across diasporic communities, members of
social movements, or moreover, to challenge ongoing research foci and public
policy. In this way, they act as alter-global discourses that are not in opposition to
globalization, yet might be used to create a different sort of globalization that is not
characterized primarily by global markets, accountability rhetoric, and national
identity formation. They also can act as tools of psychoanalytic understanding of
experience. They begin as nomadic epistemologies, yet subsequently function as
folds of interiority and exteriority, as hinges across potential apolitical and potential
highly political activity, enabling a shared discourse that coexists in more than one
parallel world of mathematics education activity.

My 'research method' for this chapter is one of philosophical inquiry as concept
construction (Deleuze 1996). This is used to make a synthetic analysis of a collection of community-based, participatory research projects. My style of engagement with my readers is one of invitation, which constructs a potential opportunity
for transferability of the generated theoretical framework and requisite structures of
collaboration. I ask you to imagine a 'we' as mathematics educators, trying out such
concepts that can exist at the same time both in mainstream mathematics education
and in ways that do not require common assumptions of mainstream mathematics
education. If this invitation to a 'mind experiment' can help 'us' to explore such
ways of working as mathematics education researchers and practitioners, then 'we'
can also consider new, future sets of concepts that would serve a similar function. In
this manner, the ways that 'we' as mathematics education scholars use the ideas
of work/labor/action, disparity/desire, youth-leadership/voice/participation, and
architect/scientist/artist to conceive our work are illustrative models of a more
general approach to existing in multiple worlds. We live in traditional school

mathematics environments where we use typical categories of analysis to understand our work, and also in these same environments, altered through new discourses, to reveal different expectations, outcomes, opportunities, and boundaries of possibility. The same discourses further open up possibilities of working in new ways, as members of a broader community of social activists, as members of political parties, as affiliated members of diasporic communities working for human rights, and so on. The discourses function as a kind of intellectual hinge or fold, so that 'we' are always both educators and active or passive political actors associated with a variety of other groups, institutions, and communities, with unique perspectives that can be articulated or not as we live and work in these multiple worlds (Appelbaum 2017). Who are we becoming as we live and work in these multiple worlds? We are always becoming our 'new' selves, and these unfolding subjectivities are complicated and always changing with respect to differences of intention, interest, focus, forms of knowledge, ways of working, and so on (Appelbaum 2017). The differences that might be taken to characterize our subjectivities are, in the words of Todd (1997), related to disparities and desires, norms, hopes and fears; and in this sense these epistemological differences establish in their folds and unfolding, that is, in our uses of them, an alterglobal, topological space of mathematics education.

Because of the ways that we are integrating epistemology with our topologies of mathematics education, we can recognize a kind of agency on our own part. That is, by changing ourselves, through our epistemologies, and then the uses of these epistemologies, to construct folds and unfolding that establish topological spaces of mathematics education, we might significantly change the world of mathematics education. Any act of theorizing, reflection, research, etc., creates an agency associated with subjectivity, as described above. The fold is in one sense the name for this relation to oneself as a subject. The fold is the effect of the self on the self, created by the act of looking at oneself, becoming outside of oneself, who is already inside. In this way, we can see that a sociopolitical agenda can benefit from a 'view from the fold', because any political struggle is going to necessitate a new form of subjectivity, that is, a new set of unfoldings.

Which set of categories do we take as the core folds for the moment? Those we notice creates a particular subjectivity. This also could be said to define a scale of observation. Are we at a micro level of a student in a classroom, or at a more macro level of the effects of mathematics curriculum policy on the patterns of refugee migration globally[1]? Are we studying teachers in schools or members of social groups learning the skills of statistical analysis in order to effect change in labor laws? I suggest that the more powerful folds are related to those nomadic sets of terms that can be applied at the most variety of scales, and I further suggest that the

---

[1]This might be taken as a self-mocking tone: who among us is so self-important as a mathematics educator to imagine that our work might have such a significant impact? On the other hand, how would it change the nature of our field if we took our work so seriously and thought about the potential to make an impact on such important crises as refugee migration, climate disasters, and so on, and worked to make this a reality?

three examples in this chapter are good ones for this very reason. These potentially powerful folds come to be 'creases' in our fabric of mathematics-education/ mathematics-education-scholarship; through repeated folding and unfolding of interiority and exteriority, the 'crease' is what I understand as the lasting trace of such repeated folding. The more that researchers consistently carry out the same foldings, this ongoing construction of subjectivity, this application of nomadic epistemologies to change oneself, in order to change the world, takes on a "character", analogous to those creases in a person's face that emerge over time along with their life history and personal commitments. What this means is that one's work as a mathematics educator is intimately connected to one's social, cultural, political and ethical commitments. If one hopes to improve the prospects of oppressed communities, or to rescue the planet from the devastation of irreversible climate change, or to make one's professional work consonant with one's views of privilege and disparity, then the character of one's creases are increasingly important.

3.1 Multiple Representations in the Classroom

As an example of how these discourses can be used to reframe a typical classroom context, I work with teacher education students and teachers to plan lessons and projects that meet school expectations. We know that our activity must take place mostly in classrooms in the school, and result in the achievement of content objectives. We also consider how the effort in the room is perceived by the young students: We design the lessons so that activity contributes to the formation of a community with a purpose. We plan activities that will enable each member of the class to exhibit leadership. Our planning and assessment judges whether every opportunity is taken to use materials to model mathematical concepts in order to convince others of one's ideas, to better analyze or explain mathematical relationships, and to raise new questions not yet explored. In this way, we are applying the various discourses discussed in this chapter to plan and assess the ongoing work in school. Beyond this basic application of the discourse, and the reflection on how this discourse changes our understanding of what is happening in the classroom, we also plan the curriculum to be grounded in social justice action projects with community groups. We the teacher-students, and young learners, together learn the mathematics we need to know in order to help these groups attain their goals for themselves. Learners organize and facilitate the class meetings, in order to experience varying styles of leadership and participation. Manipulative materials, graphical representations, and physical structures built to scale, are incorporated in projects, in order for students to experience the use of particular models chosen for the audience they have in mind, for the persuasive or analytic purpose they need, or for the power to impact on others.

In this example, the teacher education students are working as the 'teachers' who need to document for the school how their students are meeting a city-wide set of objectives for every six-week period of a school year. This city-wide curriculum is consistent with the state standards and the national core curriculum. They maintain an assessment portfolio of evidence that they have from various methods of assessment that the students are meeting these goals. Sections of this portfolio maintained by the teacher-students also include parallel documentation for our own purposes, addressing the ways that the learners are taking action in the sense of Arendt more often than working or laboring. Other sections of the portfolio contain records of the ways that each young learner is participating in the group activity, and the effects of their own choices on the ways that the young learners can experience leadership, be heard by others so that they recognize themselves as having voice in the classroom, and participate in ways that support the leadership of others, as describe by Pitt. The teacher-students further keep records of the ways that each young learner is using models of mathematical concepts, and how their own choices of models have helped each young learner to use models in each of the three forms outlined by McElheny.

3.2 Youth Mathematician Laureate Project (http://yomap.org)

The Youth Mathematician Laureate Project works outside of formal school programs to encourage civic governments, non-profit agencies, and other organizations and institutions to support young people in using mathematics to build communities. Whether their terms are one year, six months, or otherwise, young people are identified through criteria proposed by the project, and supplied with workspace and resources. Some youth, honored with the title "laureate", create a program of projects with various agencies and organizations; others facilitate community gardens, carry out investigations on behalf of their local watchdog associations or governing councils, or meet with groups of various ages and interests to help them become more independent in their own mathematical practices. Still others work with development agencies, NGOs, and community organizers, to help regional communities plan sustainable forms of land use that enable the creation of thriving infrastructures and indigenous cultural celebrations. As the project website describes,

- Youth mathematician laureates are recognized for their ongoing accomplishments with further platforms that support the expanded use of mathematics in socially compelling and aesthetically impactful ways. Laureates work with groups to listen. Laureates work with groups to learn. Laureates work with groups to create. Laureates do it again, and then again, together with others.
- Mathematician laureates are rooted in principles of accessibility, inclusion, self-determination, collaboration, sustainability and life-long learning. In all of their work, they explore the stories behind the community experience, and they believe in the power of mathematics to express what's most important.

Mathematician laureates value community conversations and hands-on creation
—and the deep understanding that can come from the combination of the two.

- Together with others, mathematician laureates get their hands dirty. They work
  with reclaimed and recycled materials and tackle large-scale projects to show
  what's possible when a group of creative and dedicated people set out to make
  or do something wonderful.
- Mathematician laureates work with others to tap into intellectual and creative
  spirits. They teach in public schools and around a table, in gardens, and on the
  streets, in parks and swimming pools, on bridges and inside caves, making the
  connections between understanding, new ideas and hands-on making. And
  every time they teach, they learn something new that adds to the story.
  Mathematician laureates support the ongoing development of a community of
  creative people who have this experience and activate it across issues and dis-
  ciplines. They come to know that this learning is what brings us all together as
  an ever-expanding community, and that it is this way of learning and creating
  together that helps us make things better in our neighborhoods, cities, regions,
  and throughout the world.

The Youth Mathematician Laureate Project is at once an approach and a community. The laureates are teachers, makers, organizers, leaders, students, advocates,
dancers, performers and everything in between, working with mathematics to enact
creative civic leadership.

The Laureate project is an excellent example of a way of working advocated in
this chapter. Because it is situated outside of school programs, it is free to invent its
own concepts of mathematics education. Instead of designing ways to 'teach'
young people, it begins with the premise that there are many youth who are using
mathematics as an art to create communities, and that these youth can be supported
in their efforts in order to promote particular, new perspectives on mathematics and
the learning of mathematics. By definition of a laureate, the youth who are identified have on their own found ways to use mathematics to support Arendt's concept
of action that builds community. They have used mathematics as a tool of leadership, as conceived by Pitt, to promote participation of others in action. Whether
they are using models in the various ways is not clear unless we examine their work
more closely; perhaps this third discourse of models from McElheny might be
incorporated more fully into the ways that youth are identified and the ways that
they are mentored in project development. Nevertheless, from the work of the youth
mathematician laureate project, we can move into school programs and ask when
and how they might take on some of the same characteristics. Might
community-building projects be more present in school programs? Might the action
projects that laureates initiate be models of projects that can be carried out by
school groups as part of the curriculum? I have written elsewhere of children
researching community problems and presenting recommendations to local governments, redesigning school spaces, and working to support public writing projects
in mathematics (Appelbaum 2009b); I have also written of secondary students
exploring the role of mathematics in democracy, and the potential of mathematical

practices to manipulate voters into false choices (Appelbaum 2009a, b). These and other school-based activities would in this way be pulling the experiences from outside of school back in, while the projects of the laureates outside of school are using what they have been learning in school. Laureates further invent their own methods of supporting mathematics learning outside of school.

4 Reframing Purpose and Action

The three discourses are not meant as prescriptions for researchers or teachers, nor as solutions to problems, but rather as heuristic examples of how coexisting discourses can help us to understand our own work in new ways, and in the process, change, for others, and ourselves, the worlds of mathematics education in which we participate. In the brief examples, work in school led to interaction outside of school, and work outside of school had the potential be brought back into school. This would effectively unfold through the use of any alternative discourses with which one might care to experiment. Because the discourses are/can coexist in parallel worlds, they allow us to work in the common sense world of accountability and clearly stated content objectives, and at the same time, to work outside of that world, using language that works in both. We can support the experiences of teachers and students, curriculum designers, and researchers, and also challenge those experiences, at the same time. We can work at once inside of school and outside of schools. We can be alterglobal, because we are addressing political and social issues within the experiences of learning mathematics, rather than by expecting mathematics to later solve social problems. We can remake ourselves, and, in turn, remake our worlds of mathematics education. This chapter is an invitation to readers to create their own nomadic terms, as part of this tactic of alterglobal mathematics education, as a solution to the recurrent boundaries between mathematics education and social change practices. There is, in a sense, an 'algorithm' that can be followed for this sort of political change effort:

- Change our words…
- Which exist in common and also in a different world of social justice and political change…
- Which leads us to see mathematics education in places outside of school…
- And then to bring the same ways of talking and working back into school…
- So that we can now find a new kind of mathematics education in and out of school…

Mathematics education is not a panacea for all international crises; but there is a long tradition of working locally in the context of global forms of injustice and alter-global movements for change. Mathematics education can be an important player in these "glocal" (Pleyers 2010; Tarc 2013) efforts. There remains the danger of working locally on issues that transcend the local without impacting on the global through the work—of using the local as an oasis away from the global. But surely this is yet another topic for further discussion—the reader is invited, yet further beyond the creation of new, nomadic epistemological terms, to struggle with the local-global as a dichotomy that needs further nomadic transformation itself, perhaps through mathematics education innovations.

5 Playing with and Around the Unconscious

Mathematics education is in a state of never-ending and escalating stress: repeating old experiences; calling for more connections to everyday life; demanding more rigor, or less attention to standard algorithms, or calling for more attention to problems by types; pursuing homogeneous—or heterogeneous grouping; proclaiming a need for careful attention to teacher questions, or for teachers to avoid questioning; and on and on we go. We do not recall the prototype of our actions, but instead see each moment of research and practice as ever-newer opportunities to jump in—to jump in, in the same way, again and again and again, writing new curriculum materials, training teachers in seemingly new ways, blaming the tests of achievement, designing new tests of achievement, returning or leaving the need to speak or writing more in mathematics classes, returning or leaving the need to place students in community situations where they use their mathematics skills and concepts to help groups of people to reach their visions of a better life ... These new moments capture with an ever-new certainty the stresses and challenges that teachers face, the ever-new needs of an increasingly mathematized society, teachers who are blamed for not teaching what students need to know, students who are blamed for not working hard enough, not being committed enough, not being capable enough, curricula blamed for disconnecting from the everyday life of students, curricula critiqued for dwelling so much in the everyday life of students that they never concretize of abstract or generalize or specialize with enough rigor to advance into other realms of mathematical sophistication, or in order to truly develop calculation skills, or ...More blame. More control. More need to predict, to cure, to comfort, to welcome, to accommodate, to promote ... what Lundin (2012) has described as "a vain struggle to bring about the perfect delivery which will teach men and women to 'think mathematically' and thus finally give them means to understand and control the social and physical reality" (p. 84).

As authors as varied as Fox (1993), Phillips (1998), and Brown (1973) have discussed, the most important things are really difficult to teach, the most important things seem to be learned without being taught, and the knowledge of what is being learned is lost and hidden to both teacher and student at most all times. No one can know beforehand what is of personal significance, we cannot control what is 'learned' because all moments in time are educative, mis-educative, non-educative, and create as much ignorance as knowledge. Most efforts to measure 'success' in some way—test scores, reduction in bias, political revolution, higher states of consciousness, a defined notion of a better way of being alive—seems to always lead to disappointment (Taubman 2012, p. 182).

What are we learning, about ourselves, the world, mathematics, mathematics education, and so on, when we use nomadic epistemological terms? This, too, is unclear and, to use a Freudian concept, forever a 'purloined letter', always further away from us the closer we seem to come to it. More to the point, from the phrase above, no one can be taught or learn something of significance, because no one can know … exactly what *is* of personal significance. Yet, does this mean that we wait and see what feels significant, and then pounce on it, forever ignorant of what we might have learned if we had avoided this very action? In some worlds of psychoanalysis, most significant are those very things we resist; what appears least significant may be what we have the greatest trouble admitting we want. Much of our effort in mathematics education falls into the realm of what Paula Salvio (1999) called the "narratives of cure". This trope locates questions of education in languages of disease, symptoms, and recovery, and conjures up an 'ending' to the problems faced by mathematics education practice. We enter practice worlds and find a symptom, a feeling of dis-ease, a dysfunction, and we want to find, through research or modifications of what we prescribe as a curricular encounter, a fantasy cure that solves the problem and makes school mathematics 'well'. "By clinging to the fantasy that we can indeed cure our students of ignorance," writes Salvio, "we keep from our conscious awareness as teachers our own frailties, fears, and anxieties about our competencies, intelligence, and emotional stability" (Salvio 1999, p. 187). Taubman goes further, adding that

> … the desire to cure and the need for hope, both so much a part of the therapeutic project, keep us from facing the terror of our own mortality which comes to us each second, if we give ourselves over to the shocking contingency of the moment. It is fear that freezes us and makes us want to freeze reality such that it can be measured, and sectioned into right and wrong answers, curriculum designs or final solutions. (Taubman 2012, p. 185)

In Taubman's terms, there is a powerful avoidance of personal and ethical crisis that is at the heart of the social and cultural need to plan a perfect and socially just mathematics curriculum, despite the doomed proposition of such a project. There is a paradox that makes this both a hope and a failed project from the start. The project maintains and sustains a fundamental human terror in the very act of attempting to overcome it: "… it is not only the fear of contingency that arouses the desire to cure and control," writes Taubman, "…it is not just the fear of our mortality or that we won't have what we desire or have what we don't want that makes us cling today to science as offering hope for a good or better outcome" (Taubman 2012, p. 186). Even as we take on these alternative discourses that remake ourselves, we can easily slip into the same trap—oh how nice that I am using McElhney's models or Pitt's concepts of voice, leadership and participation, so that now we can make the perfect curriculum. No! This will only maintain the terror at the heart of mathematics education.

Following Taubman, we can suggest that it is the playful refusal to latch onto any one nomadic epistemology, to wander through ever-changing discourses and never settle down, that might be a way to live in mathematics education work without hope and without the desire to 'fix' the curriculum. Who am I, as a mathematics educator? As a mathematics teacher? As a mathematics education researcher? As a mathematics education policymaker?—These are questions to be avoided, as they demand a fixation, a reification, a settling into a discourse. They are attempts to resist the confrontation with the terror, yet feed the terror at the same time: "... the terror," writes Taubman,

> ... that the very kernel of our being, the inner real me', is in fact not mine at all, that evokes our desire to control and cure. It is the fear that the painful reality that greets us may in fact be an answer to our own desires and questions. It is the terror that the 'I' we hear in our innermost thoughts and the 'me' of our conversations may consist of fossilized answers to what we imagined were others' desires. ...This terror at the knowledge of and from the unconscious can provoke the defensive rush to cure or control others. Hope defends against such terror because it offers and undisturbed tomorrow. (Taubman 2012, p. 186)

If we do focus on something other than outcomes that construct this kind of a doomed 'hope', we might engage with what Taubman called the swirling confusions of our own existence. We would use the nomadic discourses to do something other than work toward an assumed end. We would avoid the fantasy of 'a socially just, politically empowering, equitable and technologically rich mathematics education suitable to the crises of a global consumer culture fraught with population shifts, politically independent mega-corporations, and racially and ethnically loaded injustices'. This entails opening ourselves via new alternative discourses that coexist with, yet challenge, the proverbial cure and optimism.... in dialogue with the realities and daily details of mathematics education life in school, in communities, at home, in families, temples, churches, mosques, social agencies, parks and forests and shopping centers, in prisons and foster homes, and under bridges where those without homes congregate, and on the corners of privilege, in the gardens of those struggling to survive famine. Teaching in this sense becomes re-symbolizing and re-constructing alternative futures; teaching mathematics becomes a form and rhetorical structure for these acts of re-symbolizing and re-constructing. Mathematics education would be found in places where this happens. Those places might likely be outside of schools, but can also take shape in classrooms where those adults responsible for facilitating activity are exploring the potential for their teaching to support this kind of work, who playfully shift across nomadic epistemologies and experience the unfolding world of mathematics in their classrooms communities in these newly shifting ways. Those places would be where those responsible for the orchestration of activity ask,

> How can mathematics education work with, within, and in parallel to crowds of recognition, to protest dispossession through performative, trans-national projects? How can we use commitment to alterglobal action in supporting complicated conversations that are recognized and that recognize? How can we nurture maker-communities of conviviality, designing, and committed aesthetics? How does mathematics education strengthen recognition and when does it weaken dispossession?

Our response might be, 'Perhaps when we use the ideas of work, labor and action to reflect on practice, plan activities, and structure collaborations. Maybe when we use models in at least three different modes. Perhaps when we use the ideas of youth voice, leadership and participation to reflect on practice, plan activities and structure collaborations. Maybe when we use other nomadic categories and discourses.'

6 Holding onto Knowledge

As we place the nomadic remaking of ourselves into a global context, and consider the relationships among mathematics, mathematics education, and trans-national social and cultural crises, the aftermath of political and military confrontations, and the dynamics of diaspora communities in an ever-changing corporate consumer culture, we might wonder about the location of mathematics itself, and the teaching and learning of mathematics, in its relative importance to other aspects of mathematics education. In this way, we might imagine shifting the center of our nomadic languages away from mathematics skills and concepts, toward terms that are wholly independent of the discipline of mathematics. Hence the seduction of 'voice, participation and leadership', or 'work, labor and action'. Even McElheny's models are independent of a mathematician's usual thinking about modeling in mathematics. That was of course the point of these nomadic epistemologies, to be independent of the usual mathematics education discourses while also working alongside them. Yet, must we think that the most powerful terms come from the alterglobal crises? Paul Tarc argues for maintaining what he calls a "knowledge approach" when considering education in international contexts, and especially given trends to teach for dispositions and fixed outcomes in what he labels our "'making a difference' moment" (Tarc 2013, p. 114). Tarc is writing from the perspective of a student studying abroad from their home, and is suggesting that even in this context, it is not wise to focus on lofty social justice goals, global citizenship skills, designing a 'transformative experience', 'experiencing poverty', and so on. Instead, like a good storyteller who never explicitly states the moral of the story, but supports the audience's construction of their own significant meaning in light of how the storyteller crafts the nuances of the story, "...the focus may be directed ... to knowing more and understanding more deeply the relations between the world, others' lives and one's not always so stable or rational self" (Tarc 2013, p. 114). Bringing his ideas into mathematics education, and looking at how mathematics education might contribute to an alterglobal curriculum, we can use Tarc's three levels of knowledge to ask that a mathematics curriculum: More explicitly foster a richer understanding in a traditional sense of other societies and cultures, of alterglobal issues such as global techno-cultural trends and disempowering consumer culture, the migration patterns of economic and political refugees, and so on; Enable learners of mathematics to learn across difference; And support learners attunement to their own affective responses signaling inner desires and psychical dynamics shaping and sometimes shutting down meaning making. What this demonstrates is the potential of this mode of re-thinking mathematics education to lead us into new discourses that have no initial connection to the discipline of mathematics. It is beyond the scope of this chapter to elaborate on yet one more set of three categories that we can use to plan, assess, design, and alter mathematics education events. It is important nevertheless to acknowledge the infancy of such a grand project as a nomadic, alterglobal, social change mathematics education through this kind of example of the rich potential for exploring such discourses. There is no clear 'point' of this inchoate alterglobal mathematics curriculum that would function as the hope that maintains the problems. We would be replacing this designed hope with processes of transformation in becoming ourselves, and thus our worlds, by developing strategies and tactics for confronting difficult knowledge. There are some places where traditional school mathematics is considered 'difficult knowledge'. In this work, though, what we mean is that we start—now—thinking of mathematics education experiences as already constructing languages and frameworks for discussing difficulties in learning. Tarc evokes the example of a student who 'wants' to be opposed to specific forms of oppression or injustice, yet is also confronted with their own "implicatedness" to the oppression" (Tarc 2013 p. 115). How has mathematics education helped such a student previously? How can mathematics and mathematics education help them in ways that it has not yet? Alter-global social movements provide nomadic epistemologies for thinking through these kinds of questions by placing mathematics education in the context or broad social justice movements. The nomadic discourses link particular mathematics skills and concepts to their potential to become frameworks for recognizing and fostering human recognition and dignity, and structure relationships that evolve over time when one builds networks of educators doing this sort of work. In this sense, we return rather explicitly to the value of particular mathematical skills and concepts as learned, elaborated upon, and practiced over time as part of an education in and out of school structures.

7 Engaging Coloniality

One particularly important direction for this new phase of mathematics education is confrontation with the invisible ignorances that structure our worlds of possibility. It is usually a vicious and especially strong resistance that manifests itself as reality, and in the case of mathematics, one version of this resistance is the legacy of mathematics as an epistemological as well as viscerally experienced tool of colonialism. Mathematics as defined by most mathematics educators is a narrowly conceived version of an apolitical, culture-free, rational form of pure knowledge, hiding the very political, culturally-embedded, paradox-rich, and often nonsensical European discipline that has come to be the definition of the discipline. School mathematics is confidently codified as consistent with this narrow view, despite the important work in ethnomathematics and the history of education's complicit roles in colonialism. So, despite numerous examples from ethnomathematics countering the hegemonic common sense practices of mathematics education, the dominant mode of school mathematics and popular discourse around the discipline worldwide obliterates the history and power of non-Eurocentric versions of mathematics. Here is another rich direction for further research. It is beyond the scope of this chapter to begin to pull into the discussion the enormous and extensive literature of ethnomathematics. What is important, however, is to recognize that we are living the legacy of colonialism, those situations in the world in which a colonial power has set in place a colonial administration that establishes a hierarchy and control. This legacy is sometimes referred to as 'coloniality'. Colonial administrations have almost been eradicated from the capitalist world-system, yet there persists a cultural, political, sexual, spiritual, epistemic and economic oppression/exploitation of subordinate racialized/ethnic groups by dominant racialized/ethnic groups with or without the existence of colonial administrations. The persistent assumption of mathematics as a 'universal language' is at the core of this alterglobal project. The myth of a neutral, universal mathematics is responsible in important ways for a mythology about the 'decolonization of the world'. Mathematics can be blamed for obscuring the continuities between the colonial past and current global colonial/racial hierarchies. Mathematics contributes to the invisibility of 'coloniality' today. In this way we identify yet another important set of nomadic, alterglobal categories for further exploration: Coloniality is a lingering memory of colonialism, and it is also much more. Particularly in its active construction of ignorance around alternative knowledges and ways of being, coloniality is a pervasive force for which we cannot blame particular individual people or institutions anymore, yet through which ways of thinking and being are rendered non-existent. As an intellectual field of scholarship and practice, it helps us to find alternative epistemologies and social structures for this nomadic project of remaking ourselves. Recognition of coloniality makes it even more imperative to take on the project of remaking ourselves in ways that recognize and honor these knowledges, and in ways that accord these knowledges their dignity without appropriating them in a new form of coloniality.

One clear continuity of coloniality is the perpetuation of only one narrow form of mathematics and mathematical practices as a mythologized, culture-free, politically neutral discipline, against the grain of alternatives that abound around the world. Mathematics educators tend to see such a situation as something to appreciate, yet as something far beyond the scope of their own obligations and actions. Mathematics educators are left powerless in the wake of this global coloniality, which is experienced as far bigger than any one group or network of researchers or teachers or policy-makers, a reality to accept as we move forward. Yet, if not us, as we remake ourselves, then who will begin the important work of rethinking the very nature of mathematics itself? Paraskeva (2016) writes,

> Coloniality is the memory, the legacy of the colonialism, yet it continues to be reborn through neoliberal hegemony as a pervasive colonial power that has strong epistemological ties. So much so that we nay not perceive its power; instead, see our position as one without agency, never realizing how much of the world we do not understand because we lack the language and knowledge. (Paraskeva 2016, p. 57)

Coloniality is in this sense the ignorance that allows us to create knowledge, albeit knowledge that reproduces coloniality itself.

The point is not to correct or counter coloniality, to improve education with new outcomes or goals, for that would be a re-inscription of hope and the fears that perpetuate the very need for hope. Any notion of progress that takes this form is hardly useful. More helpful would be to appreciate how the notion of knowledge and progress that we have inherited operates mostly as a mirror that distorts what it reflects. What we find in this mirror is not necessarily just a fantasy of pure imagination; there are realities that are reflected in the myriad forms of research and practice which are found to be significant by many mathematics educators. The tragedy of our situation, if we might call it such, is that we have been guided by this mirror to accept the images in the mirror, and to accept these images as our own. In this way, we continue to be what we are not, or perhaps what we might not want to be if we only had the chance to think about it. … Builders of coloniality? … Fearful of the inability to establish progress simply because we continue the fantasy. Yet this is only a problem of closure if we think that that the end is near, that we are supposed to solve things in this kind of curative, hopeful design process. Instead, with the remaking of ourselves to see the world in new ways, we also see that we do not need to carry along the coloniality of closure either. We can find new translations across nomadic discourses of possibility, and use them for now to form alterglobal affiliations across differences, and work together in action, as Arendt would suggest, to create communities. These communities are unhindered by national or disciplinary or regional or other boundaries, and instead form and reform as needed.

8 Changing Oneself Without Becoming a Self

It has become a rather hackneyed expression that one must become the change one wants to see in the world. This is not the point of this chapter. Yes, Ghandi, the source of that expression, was truly inspiring, and I do not mean to dismiss his importance—to the world, to social change, and to the many people who have benefitted from this idea. However, I am advocating something a little different. One might say I am advocating a parallel, independent project that can coexist with Ghandi's, and which I believe could contribute as much to social change without falling prey to the fantasy of knowing what is best and how to achieve it. There are many apocryphal stories in and out of mathematics education and revolutionary practices for social justice claiming that means and ends arguments have derailed the very movement, or worrying that those working with hope have put themselves and others in danger. Taubman would possibly see these stories as examples of hope's best hope, so to speak. What is powerful in the nomadic epistemology approach advocated here is the tentative and shifting nature of the alterglobal movements, which do not demand a fixed identity or unfaltering commitment to a universal truth, and instead enable action and community in the moments of change and fluidity. It is this power of the alterglobal that we can connect with in mathematics education.

References

Appelbaum, P. (2009a). Against sense & representation: Researchers as undetectives. *Educational Insights: Journal of the Centre for Cross-Faculty Inquiry, 13*(1). http://www.ccfi.educ.ubc.ca/publication/insights/v13n01/articles/appelbaum/index.html.LastvisitedDecember8,2017

Appelbaum, P. (2009b). Taking action—Mathematics curricular organization for effective teaching and learning. *For the Learning of Mathematics, 29*(2), 38–43.

Appelbaum, P. (2012). Mathematical practice as sculpture of utopia: Models, ignorance, and the emancipated spectator. *For the Learning of Mathematics, 32*(2), 14–19.

Appelbaum, P. (2017). Disordered order, ordered disorder: Threads, folds and artistic action. In H. Straehler-Pohl, A. Pais, & N. Bohlman (Eds.), *The disorder of mathematics education* (pp. 273–290). Switzerland: Springer.

Appelbaum, P. (undated). Youth mathematician laureate project. http://yomap.org. Last visited April 30, 2017.

Appelbaum, P., & Gerofsky, S. (2013). *Performing alterglobalization in mathematics education: Plenary in the form of a jazz standard*. Opening Plenary Address. CIEAEM (International Commission for the Study and Improvement of Mathematics Education). Turin, Italy, July 21–26.

Arendt, H. (1958). *The human condition*. Chicago: The University of Chicago Press.

Brown, S. I. (1973). Sum considerations: Mathematics and humanistic themes. *Educational Theory, 23*(3), 191–214.

Butler, J. (1997). *The psychic life of power: Theories in subjection*. Palo Alto, CA: Stanford University Press.

Butler, J. (2010). *Frames of war: When is life grievable?* New York: Verso.

Deleuze, G. (1996). *What is philosophy?* New York: Columbia University Press.

Deleuze, G., & Guattari, F. (1987). *A thousand plateaus: Capitalism and schizophrenia*. Minneapolis, MN: University of Minnesota Press.

Fox, M. (1993). *Radical reflections: Passionate opinions on teaching, learning and living*. Washington: Harvest Books.

Kirshner, D. (2002). Untangling teachers' diverse aspirations for student learning: A crossdisciplinary strategy for relating psychological theory to pedagogical practice. *Journal for Research in Mathematics Education, 33*(1), 46–58.

Leavy, P. (2017). *Research design: Quantitative, qualitative, mixed-methods, arts-based, and community-based participatory research approaches*. New York, NY: Guilford Press.

Lundin, S. (2012). Hating school, loving mathematics. *Educational Studies in Mathematics, 80*(1/2), 73–85.

McElheny, J. (2007). *Artists and models*. Video of March 2007. Presentation at the Museum of Modern Art, New York. http://www.gvsmedia.com/video-2/w-lFT7CCuT8/JosiahMcElheny-presenting-at-MoMA. Last visited October 2008.

Mellin-Olsen, S. (1987). *The politics of mathematics education*. New York: Kluwer Academic Publishers.

Paraskeva, J. (2016). *Curriculum epistemicide: Towards an itinerant curriculum theory*. New York, NY: Routledge.

Perry, B., & Dockett, S. (2002). Young children's access to powerful mathematical ideas. In L. English (Ed.), Handbook of international research in mathematics education (pp. 81–112). New York: Routledge.

Phillips, A. (1998). *The beast in the nursery: On curiosity and other appetites*. New York, NY: Pantheon.

Pitt, A. (2003). *The play of the personal: Psychoanalytic narratives of feminist education*. New York, NY: Peter Lang.

Pleyers, G. (2010). *Alter-Globalization: Becoming actors in a global age*. Cambridge, UK: Polity.

Salvio, P. (1999). Reading beyond narratives of cure in curriculum studies. *Journal of Curriculum Theorizing, 15*(2), 185–188.

Skovsmose, O. (1994). *Toward a philosophy of critical mathematics education*. Dordrecht, Holland: Springer.

Sliva Spitzer, J. (2003). *Teaching inclusive mathematics to special learners, K-6*. Thousand Oaks, CA: Corwin.

Star, S. (1995). *Ecologies of knowledge: Work and politics in science and technology*. Albany, NY: State University of New York Press.

Tarc, P. (2013). *International education in global times: Engaging the pedagogic*. New York, NY: Peter Lang.

Taubman, P. (2012). *Disavowed knowledge: Psychoanalysis, education, and teaching*. New York, NY: Routledge.

Todd, S. (1997). *Learning desire: Perspectives on pedagogy, culture, and the unsaid*. New York, NY: Routledge.

Tripathi, P. (2008). Developing mathematical understanding through multiple representations. *Mathematics in the Middle School, 18*(8), 438–445.

Part II
Researchers in the Sociopolitical
in Mathematics Education

Recognising and Identifying the Participant and Researcher in Mathematics Education Research: A Sociopolitical Act

Lisa Darragh

Abstract Researcher positioning is an important consideration in acknowledgement of the power inherent in the act of research. As such, the concept of identity may be deployed to consider such a political act. In this chapter, I examine the use of identity in mathematics education research to both identify the participant and also to consider the identity work of the researcher. I use Butler's (Theatre J 40:519–531, 1988) definition of identity as performative to understand the ways in which the subject is produced in discourse and incorporate the notion of identity recognition to consider the ways in which discourses are drawn from in performing and recognizing. I engage in a reflection of my own identity work as a researcher and suggest we should consider the researcher identity as being multiple, temporal, fearful and desiring, and also complicit in the (re)production and disruption of broader social and political discourses in mathematics education.

Keywords Mathematics identity · Performative identity · Researcher positioning Discourse

1 Introduction

I take as a starting position that the research act is powerful (Gutiérrez 2013; Pais and Valero 2012; Planas and Valero 2016) and therefore the role of the researcher in this act of power is due considerable attention. Over the past few years there have been a number of arguments emerging from within mathematics education about the importance of positioning the researcher within the research project. Gutiérrez, for example, states:

> To engage with the political, the [mathematics education] field needs to value and encourage researchers to position themselves within their work (e.g., articulating those

L. Darragh (✉)
Center for Advanced Research in Education (CIAE), Universidad de Chile, Santiago, Chile
e-mail: darraghlisa3@gmail.com

© Springer International Publishing AG 2018   77
M. Jurdak and R. Vithal (eds.), *Sociopolitical Dimensions of Mathematics Education*, ICME-13 Monographs, https://doi.org/10.1007/978-3-319-72610-6_5

aspects of their identities and ideologies that inform their choice of research projects, the design of such projects, the kinds of questions asked, and findings produced). (Gutiérrez 2013, p. 62)

Gutiérrez continues this line of argument in a dialogue, published in the same journal issue, in which a number of high-profile writers and editors discuss this very topic (D'Ambrosio et al. 2013). These authors in mathematics education suggested that *all* research is non-neutral, is political and is about issues of identity and power —and for this reason a visible researcher positioning is important. Similarly Valero argued the importance of considering, amongst other factors, "who we are and how we choose to engage in academic inquiry" (Valero 2004, p. 6). However, Walshaw cautions that in the doing of this type of self-reflection of researcher position we must take care not to "romanticise the self" (Walshaw 2010). Writing ourselves into the research does not by itself counter the effects of "power, privilege and perspective in the research encounter" (Walshaw 2010, p. 588). The use of the first person narrative, statements regarding data as co-constructed, and discussion about the asymmetricality of power in the research interview all indicate an attention to researcher power, yet these considerations sometimes appear to imagine researcher identity as being static and singular. In contrast, if we consider the identity work inherent in the research process, we may be able to develop a much more sophisticated consideration regarding the power of the research act.

In this chapter I argue the concept of identity provides a very useful tool in which we may consider the political act of research. This tool may be deployed in two ways: firstly, the recognition of the research participants may be considered as an act of *identification*, and secondly, the researcher engages in her own identity performance in the research process and through positioning within her research. I aim to demonstrate this through a self-reflection on some of my own recent work. I first discuss the concept of identity within mathematics education and consider its usefulness in exploring some possible implications of the political act of research. I present a description of a research participant, and then follow with a critical examination of the way in which I identified her, drawing on my doctoral research and reflections thereafter (Darragh 2014). I conclude with a consideration of my own identity performance as researcher in this process and provide some implications for theory and research practice in mathematics education.

1.1 Identity and Recognition

Identity is a concept that has proven worthwhile to those researching within mathematics education and its use has grown rapidly within the domain over the last two decades (Darragh 2016). The lived experiences and relationships that students and teachers form with mathematics learning and teaching are seen as very relevant to the mathematics education enterprise, particularly in research concerned with social and political issues. Just a few examples are Solomon and colleague's work on gender (Braathe and Solomon 2015a; Solomon et al. 2011, 2015) and Mendick's (2005) discussion of the masculinity of mathematics. Martin (2000), Stinson (2013) and Nasir (2002) all provide examples of research which consider the intersection of race and mathematics learner identity. Lerman (2009) has used identity to examine social class, Walshaw (2013) takes a psychoanalytic approach and Chronaki's (2011a, b) work is concerned with a range of social and political aspects, using and furthering identity as a lens for research. Recently identity has been directly linked to issues of power (Gutiérrez 2013) within the sociopolitical perspective on mathematics education.

Along with these authors I prefer to see identity within the sociological, initiated in the work of Mead (1913), and I draw on Judith Butler, to define identity as performative. Butler refers to gender identity, however the definition can also apply well to mathematics identity—as something first performed and secondly embodied. This means identity is constituted by a "stylized repetition of acts" (Butler 1988, p. 519). It is similar to a theatrical performance, in that the identity does not exist prior to the act; rather it is constituted in the moment. A true core self does not exist; the self is continually being made in and through performances of the self. Such performances may include style and content of speech, labels applied to the self (such as 'gifted', 'dyslexic' or 'creative') or they may be physical acts. Within the educational setting we may even see performances of the learner self through the way in which a student wears their uniform or how they sit at their desk (Youdell 2006a). In mathematics education we may witness performances such as someone saying: 'I don't have a math brain' or someone choosing to study mathematics at an advanced level.

This notion of identity appears to allow significant agency on the part of the individual, however, the performative act is actually derivative and cites an authoritative discourse. Butler draws on speech act theory to suggest "a performative is that discursive practice that enacts or produces that which it names" (Butler 1993, p. xxi). For example the term 'student' is a designation that appears to describe a pre-existing subject, yet it is the act of designation that constitutes the student (Youdell 2006a). This implies that what it means to be a mathematics student is not predefined, rather it is produced through the naming and via wider discourses (and therefore it may be produced differently). With this definition Butler adopts Foucault's notions of discourse as productive and of subjectivation together with Althusser's notion of subjection and the hailing of the individual (Youdell 2006b). In other words, identity is performed within wider discourses and the identification by someone (or by some*thing*), with authority, constitutes the individual as a subject. Butler's conception of identity has been taken up by some in mathematics education (Chronaki 2011a; Damarin and Erchick 2010; de Freitas 2008; Mendick 2017; Stinson 2008), and should also resonate with those who use Foucauldian notions of subjectivity through which to understand identity (e.g. Hossain et al. 2013; Llewellyn 2008, 2009; Walkerdine 1990, 1998; Walshaw 2001) and more broadly with those who find value in Foucauldian theories of discourse.

To unpack some of these ideas I find it useful to return to Butler's (1988) theatrical metaphor, which allows us a simpler language through which to consider aspects of the wider social and political context and discursive production of identity. I consider the roles of the *stage,* the *theatre,* available *scripts* and the role of the *audience* through which to imagine identity performances and the notions of subjection/subjectivation. The stage may be considered the local context, the theatre the wider political context, scripts are available discourses defining the limits of performance. Finally, the audience provides the crucial aspect of the social context, the necessity for the performance to 'make sense' to others (Youdell 2006a) and also includes the authoritative other or institution which may 'hail' (in Althusser's terms) the subject. In line with this metaphor, I see it as necessary to understand identity performance as simultaneously both the performance and the recognition of the performance by the audience. This means that identity does not exist within the person; rather it exists and is negotiated in the temporal and social space between the performer and another. This view of identity is useful in that it steers us away from a conception of identity as residing within an individual.

A significant limitation in the use of the concept of identity within mathematics education has been a tendency to focus on the individual (Darragh 2016). This may be due in part to the neo-liberal context in which much of this research is produced; neoliberal ideology tends to frame our thinking as primarily concerned with individual success, such as success in education (Apple 2000; Ball 2013; Harvey 2005). Mathematics education is highly implicated in this neoliberal project (Valero 2016). Research which focuses on the individual or group being identified without consideration of *who* recognizes and *how* they recognize identity performances does not provide us with the necessary tools to challenge inequity in education and use identity as a tool to radically rethink education (Youdell 2006a).

Other limitations of identity research have been wagered by some researchers within mathematics education, arguing that a focus on identity will not help to address the pressing issue of income inequality (Jurdak et al. 2016, pp. 23–24). Furthermore, identity research sometimes assumes binary between structure and agency, which "tends to locate the power to change at the level of the individual human subject, and denies this power to larger structured collectives, say institutions" (*ibid.*, p. 15). However, a poststructuralist conception of identity, such as in the work of Butler and Foucault demonstrates how selves are both subjected *by* and *of* discourses, meaning they are both constrained and *agentic* within these discourses (Hossain et al. 2013). Post-structural ideas such as these "come out of a recognition that existing structural understandings of the world, whether these focus on economic, social, ideological, or linguistic concerns, do not offer all the tools we need" (Youdell 2006a, p. 37). The identity project is to interrupt material inequality through an engagement with language and through the formation of disruptive discourses (Youdell 2006b).

Furthermore, power is not only concerned with large scale domination and oppression. Power can exert itself in much more localized ways and impact on a person in such a way that it appears to become absorbed into the very self:

> Power acts on the subject in at least two ways: first, as what makes the subject possible, the condition of its possibility and its formative occasion, and second, as what is taken up and reiterated in the subject's 'own' acting. (Butler 1997, p. 14)

This absorption of discourse by the individual highlights the ways in which the recognition of identity, or the process of identification, enacts power. With this we return to discussion of the key role of the audience in the recognition.

There are a variety of places within mathematics education in which we may see the power of identity recognition. For example, teachers recognize their students as being particular types of learners and initially this recognition may be based on appearance (skin color, clothing, mannerisms, accent or language), or perhaps based on mathematics results in a school placement test. The effects of such recognition are well documented in the research on teacher expectations (Rubie-Davies 2006; Straehler-Pohl et al. 2014; Turner and Rubie-Davies 2015). Schools recognize the students through such assessments and wield power with streaming and labeling practices, identifying and placing students in differing ability groups (see for e.g., Boaler et al. 2000; William et al. 2004; Zevenbergen 2003). Teachers are recognized through their students' test results, and identified correspondingly as effective teachers or otherwise (Robertson 2012; Zeichner 2010). They may be further recognized through playground and parental gossip. The media recognizes the mathematically able as nerds or geniuses (Mendick et al. 2008). The Organization for Economic Cooperation and Development (OECD) recognize entire countries through highly publicized, and under-problematized, testing regimes (Kanes et al. 2014) which all involve mathematics. Finally, our own self recognition is also a powerful act, in which we play audience to our own performances.

As researchers, we are another type of audience and we too wield considerable power during the investigative process. When we do research we identify our research subjects; perhaps as particular types of mathematics learners or as a 'good' teacher, or in terms of another identity group such as race, class, or gender. Irrespective of whether our research is in the area of identity, or even whether it involves actual participants, we all identify mathematics students and teachers in particular ways according to our ideological frames, our background, our contexts, our experiences, and our purposes for research. As researchers it is *our* social and political context that impacts on the way in which we recognize our research subjects (see also Llewellyn 2016; Valero 2004).

2 Recognising the Participant

For the purpose of this chapter I draw on my doctoral research (Darragh 2014) into the identity performances of learners of mathematics as they made the transition from primary education to secondary schooling. Twenty-two students were involved in the study; I met them when they were 12–13 years old and followed their experiences in school mathematics over a period of almost two years. I observed these students in mathematics classes and interviewed each of them on four separate occasions as well as interviewing their teachers and some of their parents. The 'data' I share here is an excerpt based on a reflection about a participant that I made around a year after the completion of my doctorate:

> Emily is a Samoan girl, extremely high achieving in all subjects: English, physical education, science, and mathematics included. She was delightful to interview, possibly in part because her elder sister was doing graduate work in education and she knew the research game and how to perform it. She gave long and thoughtful responses to me in interview. She regaled me with anecdotes of her engagement in mathematics games and puzzles and shared with me a passion for the subject. I found myself especially interested in her future story because the 'data' often says that both girls and Pasifika[1] students, such as those from Samoa, are under-achievers in mathematics. She was obviously an exception. I wondered if as a 'typically marginalized' member of the mathematics learner community she would eventually experience schooling in a more negative way. At one point I even had to ask myself if I was in part hoping for this so that I could write about it in my thesis! On the other hand, I was also hoping for the opposite, that her experiences would provide a counterpoint to typical stereotypes. However, it was a revelation when I realized that actually Emily never, not once in four interviews over a period of more than 18 months, discussed being Samoan or being girl as in any way significant for her experiences of learning mathematics. (Excerpt from research journal, August 2015)

In particular, I marked Emily as 'able', as 'girl' and as 'Samoan'. However, these were not identity markers that she applied to herself during our conversations about mathematics. In order to gain understanding into Emily's identity performances with respect to mathematics learning, I spoke with her a number of times about relevant experiences in the past as well as her plans for the future. Ultimately it was I, the researcher, who analyzed and summarized these conversations and who applied identity labels. I had no relationship with Emily prior to her being a part of my research project, however she was recognizable to me, in the sense that I had taught students of her age for a number of years. I taught in a community demographically similar to hers, and in Emily I recognized some of my past students. The identifications I made of Emily are not innocent nor neutral descriptions simply adding background to the research text; they are identifications produced in wider discourses, contextually based, and they have effect. In order to deconstruct my labeling of Emily I divide the description into two parts. Firstly 'Samoan girl' and secondly 'mathematically able'.

2.1 A Samoan Girl

One way in which I recognized and identified Emily was as a Samoan girl, and when I chose to categorise her in this way I drew upon wider discourses of gender and of ethnicity. In other words, I chose to describe Emily in terms of these

---

[1]*Pasifika* refers to people from the Pacific Islands, such as Samoa.

particular identity markers. The questions at stake here are *why* I chose such categorizations? Why were these discourses accessible to me in my own particular context?

In New Zealand/*Aotearoa* data is collected and statistics are presented predominantly in relation to gender or ethnicity.[2] Almost any demographical statistic (such as levels of employment or educational attainment) is immediately and always separated into male and female categories, and also classified by ethnicity. Ethnicity is categorised as Māori, Peoples from the Pacific Islands (or Pasifika), Asian, and European (or NZ-European/*Pākehā*).[3] Such educational data tend to demonstrate the underachievement of Māori and Pasifika students and show the Asian and Pākehā students to be achieving at higher levels than the overall average. In most educational statistics girls outperform the boys, but this trend is reversed in mathematics. NZ also widely publicizes results obtained in international tests such as PISA. The New Zealand summaries look again at gender and ethnicity, showing ever widening gender gaps in mathematics (Vale et al. 2016) and a maintained underachievement for those marginalized ethnic groups. In these published summaries, socio-economic status (SES) also features, for example illustrating a rapidly widening gap between wealthy and poor students (Ell, cited in Davison 2013).

My own experiences as a teacher and with knowledge of the differing experiences of different Pasifika peoples, led me to move beyond a general classification of 'Pasifika' and attend instead to 'Samoan'. But I should note that Emily gave a very subtle performance of this identity; it was not overtly stated, rather given through talk about her wider family (including those still living in Samoa) and her involvement with the church. Emily's performance of girl, by contrast was necessarily less subtle. Her wearing of a school uniform may have hidden her SES, yet it identified her gender. Indeed, schools may argue the advantages of uniform in masking SES, however it simultaneously forces students to enact one of a binary gender identity. In non-uniform schools, students may be free to perform other gender identities, but issues of poverty could be made more apparent. However, on Emily's part, her performances of Samoan and girl were not deliberate or conscious in the interview context but they were nevertheless easily recognizable by me when I first met her. My reasons for identifying and classifying her in these ways are due in part to the typical discourses of New Zealand educational academia and government.

Given that my recognition of Emily as a Samoan girl was constructed within wider discourses in the New Zealand education system, how might I have recognized and identified her if I had met her in a different country context? Currently I am living and researching in Chile and it is very apparent that the academic and social context of this country would have led to a different recognition of Emily had

---

[2]See for example http://www.stats.govt.nz/browse_for_stats/snapshots-of-nz/nz-in-profile-2015. aspx, http://www.education.govt.nz/assets/Documents/Ministry/Publications/MinEdNZEducation Profiles.pdf.

[3]Sometimes an "other" category is included, representing approximately 1% of the population.

I studied her in this context instead. Firstly, I would have defined her in terms of her social class. Social class in Chile may be seen in a surname, an address and, perhaps most immediately, in the school attended. Upper, middle and lower social class correspond with elite and private schools, subsidized schools, or charity and public schools (Valenzuela et al. 2014). Secondly I may have considered Emily's 'vulnerability' status as another form of identification common within Chilean education. Vulnerable is a label given to at-risk students based on a complex calculation including parents' overall income, educational levels, and health-risks (Ávalos-Salamé and Thomas Ponce 2007). Whilst social class may be considered a contextual factor, outside of the individual, the term *vulnerable* is very much of the body. In this way vulnerable is an identity which may be embodied similarly to gender and ethnicity (and mathematics ability).

In Chile class is tightly interwoven with race, as it is in many (if not all) other countries. The children who attend private schools are more likely to identify (and be recognized) as White than those in public schools; and public schools have much larger numbers of immigrants, usually from other parts of South America, or recently from Haiti. But in Chile race and ethnicity receive less attention. In other words the meta-narratives of social issues such as inequality in Chile use the language of social class, whilst it could alternatively, or additionally, use another discourse, or lens, such as race or ethnicity.[4] The narrative of inequality in New Zealand on the other hand is based first on gender and ethnicity whilst other identities are positioned in the background.

The dominant discourses in New Zealand and Chile have some aspects in common. Within both contexts recognition is tied to wider academic discourses; performing as an academic in any educational context requires the drawing from the specific narratives of those contexts. Secondly it is notable that these discourses are statistical; gender and ethnicity or social class and vulnerability are aspects that are counted and as such the identification is more powerful, with the power of numbers to enforce it (see Chassapis 2017 for another example of the power of the number discourse).

Giving a general description of research participants is a common practice in academia. It is difficult to find published work within education which does not identify gender, for example, be it in large quantitative studies or smaller case study reports. Llewellyn's (2008) article 'Maths with Sam and Alex' is exceptional in Llewellyn's refusal to disclose the gender of the two participants about whom she reports. The discomfort felt by the reader in not knowing demonstrates the strength of this discursive practice of classification by gender. Reporting on ethnicity or race can be more contentious. At times this type of classification is not included, and Martin (2009), for example, would argue this makes race a near invisible topic in mathematics education research. At other times race or ethnicity are mentioned

---

[4]Note, this is not to say there is no research on ethnicity issues in Chile, there *is* (see for e.g., Webb and Radcliffe 2013), rather that issues of class and socio-economic status tend to dominate the research agenda.

together with discourses of failure. Gutiérrez and Dixon-Román (2011) argue against a focus on achievement gaps between white middle-class students and African-American, Latin@, American-Indian working class students, describing how this focus contributes to deficit frameworks for these marginalized groups. Such focus on achievement gaps are, in Martin's (2009) words an "impoverished approach to race, racism, and racialized inequality" (p. 297). Alternatively research could broadcast success instead, as demonstrated in Stinson's work (see for e.g., Stinson 2013). It is clear the seemingly innocent description of the participant is problematic.

By recognizing Emily as Samoan and as girl I am adding weight to the importance of these educational discourses about gender and ethnicity. My recognition of Emily's success despite her membership in typically marginalized groups could be argued as breaking down stereotypical discourses and demonstrating the possibility of enacting identities of able *and* girl, of successful *and* Samoan. However, I suspect that the educational and statistical discourses, of girl and Pasifika, which hail Emily into being a particular type of mathematics learner only allows her to be positioned as a learner in one of two ways; *typical* or *contrasting,* and both of these positions may in fact reinforce the dominant discourse.

By denoting Emily as Samoan and as female I made assumptions of the potential for Emily to experience a negative transition to secondary school without finding out from her (or from other key people in her life) her own interpretations of her racialised and gendered experiences of mathematics education. I never asked her outright what she thought about being Samoan (/Pasifika) or girl in the learning of mathematics. Perhaps I did not want to ask a leading question, perhaps I did not feel I had the right to ask it. Yet whilst these questions did not enter the interview script, they entered the analysis in a subtler (more dangerous?) way.

> [J]ust because a student is participating in ways that we ascribe to 'successful students' does not necessarily mean that student buys into deficit notions of kids who do not participate in the same manner. Nor do students necessarily define themselves based solely on how well their behaviors or grades correlate with discourses on the achievement gap. (Gutiérrez 2013, p. 52)

It is clear to me that Emily did not define herself based on discourses of low achievement for girls or Pasifika students. Rejection of similar narratives have been documented elsewhere, for example 'Hedvig's' resistance of gender narratives during interviews (Braathe and Solomon 2015b). Yet I was thinking about the other members of these groups during the process of analysis nevertheless.

Of course Emily was recognizing and identifying me as well during our interviews. She is likely to have noted me as *Palagi,*[5] female, and possibly also saw me as a teacher. How she saw me would have affected the issues she chose to discuss. As we were both good female mathematics learners perhaps there was not a lot to say about the underachievement of girls. But on the other hand it is our ethnic

---

[5]Palagi = Samoan term to refer to European person.

differences that may have prevented conversation about this identification and her experiences related to being Samoan. I suspect that my identification drew not from sameness or difference between us; rather I identified those aspects different to the 'normal' mathematics student. That is, to be girl and to be Samoan are two ways of being 'other'. In fact, my classification of Emily as Samoan girl was possibly particularly *pleasing* to me in my role of researcher because of the opposition between this label and that of *mathematically able*. I now turn to this second identification.

2.2 An Able Mathematics Learner

My recruitment of Emily into my research project already produces her as a particular type of subject, specifically that of the mathematics learner. Firstly, this prioritises an identity of mathematics learner above any others that may be applied to or performed by the research participant. The mathematics learner is a subject construed by the discourses of mathematics education research. She is reduced down from a fully social child to a subject of analysis within terms often only of mathematics learning (Llewellyn 2016; Valero 2004). Such a view does not give due acknowledgement to the possibility (likelihood?) that mathematics may only take a very background role in Emily's life (see also Gutiérrez 2013).

Secondly, implicit in the production of the mathematics learning subject are all the varied discourses which circulate within the wider mathematics education community. These include the superiority of mathematics and necessity of mathematics within the neoliberal society, and discourses of dis/ability as being paramount to the concerns of mathematics education. Below I will unpack some of these discourses and note also the complicity of both Emily and myself in these performances.

Mathematics has long been a high status subject (see also Pais and Valero 2012). Current neoliberal discourses have served to strengthen this view of mathematics as essential for the development of a well-functioning society (economy) and as a highly desirable attribute. As researchers we enjoy the superiority of mathematics in current society. By positioning our research within this domain we may enjoy various privileges, such as a greater ability to gain funding for example. We have a number of mathematics education specific journals in which to send our articles. When we do so, "where is the mathematics?" is a question we may feel compelled to address at the forefront (for discussion see: Martin et al. 2010). Mathematics is privileged in curricula world-wide, and with this privilege comes a subtler, hidden responsibility; we must continue to promote mathematics as important, as an essential learning area for students, and therefore also an essential research domain. There are those among us who question and critique this superior position held by mathematics (D'Ambrosio et al. 2013; Llewellyn 2016; Lundin 2012; Martin et al. 2010; Pais and Valero 2012, 2014; Swanson and Black 2017). However, these researchers too, publish in mathematics education journals.

Emily was complicit in this production of herself as a mathematics learner, she knew I wanted to research mathematics learners and she performed for me her mathematics learner identity. By talking about the mathematics games and puzzles she did with her brothers in her spare time she performed for me the good student. She was 'good' in the sense of compliance (Walkerdine 1998) and good in the sense of high ability. In this ability performance she drew from another dominant discourse within mathematics education, that of dis/ability.

It was not difficult to identity Emily as an able mathematics student; despite my not asking directly about achievement in this subject (I pretended to myself as a researcher that I was not interested in this aspect). By the end of the first interview with each student, generally even within five minutes, I 'knew' which students would be considered high ability or low. This quick, judgmental recognition may appear arrogant and yet all my assumptions were confirmed later by their teachers in both primary and secondary schools and matched with later placements in streamed classes at high school and even in their end of year examination results a year later. It is not that I was perceptive, nor even correct per se, rather that I am in and of the system. My teaching experience makes me the same kind of audience member that typically wields this power in schools.

This concern with mathematical ability is demonstrative of the preoccupation western mathematics holds with mathematics ability (Boylan and Povey 2014), and the corollary views that this ability is more or less fixed. This view is highly implicated in educational practices such as ability grouping and the subsequent delivery of differing programmes of mathematics dependent on whether the said ability is high or low (Boaler et al. 2000). Such practices doubtless impact on identity performances of students (Boaler et al. 2000), and become embodied in similar ways to other identities (Jurdak et al. 2016). However, it is the preoccupation of ability discourse that frames my interest here. The dominance of this discourse leads us to construct the 'low ability' student as pathological (Walkerdine 1994). This resonates with the 'pursuit of progress' preoccupation argued by Llewellyn (2016) as evident in mathematics education research, policy, and the talk of pre-service teachers. Llewellyn demonstrates how the notion of progress "is part of the governance and (re)production generated through mathematics education research" (p. 304) and this defines and constrains the possibilities of educational research. I suggest the notion of ability likewise functions in this way.

To summarise, my identification of Emily drew from wider educational, academic, and mathematics discourses. These discourses were also somewhat context specific, the statistics discourses of the New Zealand educational system differed from those of Chile, whilst functioning in similar ways. The importance of noting the ways in which we recognise our research participants goes beyond the classification of these participants into categories and the possibly negative effects of stereotyping that might occur through this process. This recognition also serves to reinforce the discourses which helped to produce the subjects in the first place. Therefore, we need to examine closely the discourses we draw from, contribute to, and consider whether it will work against or contribute to inequities in education.

Secondly, we need to consider the ways in which this process of identification is in itself an identity performance on the part of the person doing the recognition. Clearly my own social and political context influenced my recognition of Emily, and, as discussed above, my own identity affects the way in which I may 'see' her during the process of research. But identity is not static, rather it is multiple and the performance is fluid. The writing up of research has an effect of making identity seem static; my recognition in the moment is preserved and made to seem as being forever, but of course it is not. Similarly, my researcher identity performance is also fluid and relates also to desires and ambitions, acted in the moment, as well as the temporal and geographical contexts within which I perform.

3   Performing a Researcher Identity

As an early career researcher, I am trying to find my niche within the research domain. The way I conduct research, the presentations I give and the articles I write are all performances that constitute my researcher identity. The way I recognize my research participants (including Emily) is likely to be influenced by: the type of audience I wish to address, the type of conferences I wish to attend, and the journals I wish to publish in. In other words, the type of researcher I want to be. Of course, here I must rely on recognition from my peers and hope they identify me as sufficiently academic for their journal or conference. As I worked on my thesis I was drawn to research communities such as 'Mathematics Education in Society'.[6] I felt that for my research to be 'good' (meaning both 'moral' and 'high quality') it would need to address wider issues, such as equity and access. It was in part my naïve understandings of the ways in which to be social and political in research that led me to zero in on gender and ethnicity when I cast the identifying gaze upon Emily. In doing so I may have been trying to force a story of marginalization where it did not fit. Perhaps I was looking in the wrong place; perhaps I was looking in the wrong way. Need I even have looked? My recognitions of Emily related therefore to my own performance of an academic identity and my imagined audience. The type of researcher I desired to be was enacted through my recognition of Emily.

Walshaw (2010) talks of how our desires are caught up in the research process. She goes further than a discussion of reflexivity and researcher positioning, arguing a researcher's sense of self is not only complex but continually changing in relation to the research participant. Walshaw furthermore questions how desires and fantasies map onto the researcher's sense of self. Walshaw's (2010) reactions to 'Rachel' resonate with mine to Emily, in particular her attempt to "confront, rather than slide over, the delicate issue of emotion in the research process" (p. 592). I certainly experienced emotions in the research process and these relate to future ambitions, but also to desires stemming from my past.

---

[6]See: http://www.mescommunity.info/.

I came to research the experiences of transition to secondary school from a teaching background within primary school and I cannot deny that a small part of me expected and wished to see some of my research participants beginning to experience failure and develop negative identities regarding their mathematics education. I certainly wanted and expected to see a change. Had Emily turned her back completely on mathematics once she moved to secondary school, having been such an avid mathematics student at primary school, I would have been able to embark on a critical onslaught of her secondary school mathematics experiences. This would have pleased my primary teacher self. Typically research in mathematics education tends to create a 'victory narrative' (Brown et al. 2016); the story I was keen to write would cast primary mathematics education in the role of the unsung hero.

However, regardless of whether my desire was to write positive or negative results, I was driven to provide 'understandings' about mathematics education ultimately in order to 'improve' it:

> References to such discourses seem often to shape the activity of aspirational individual researchers. The superlatives used in the construction of these narratives, however, can sometimes disguise the differences between the multiply directed motivations of mathematics education researchers (e.g., for ethical practices, to understand more deeply, to disrupt or think differently) and the operational motives that guide their actions (e.g., securing funding, getting published, recalibrating practice, working towards a Ph.D., etc.). (Brown et al. 2016, p. 288)

The operational motives of my research were more than obtaining a Ph.D.; I was also aware of the need to publish. In this, Emily was my 'money maker'. I have used data from her interviews on a number of occasions. I referred to her specifically in a journal article where I examined different ways of performing a 'good at mathematics' identity and lack of recognition of oneself in this performance (Darragh 2015b); here Emily[7] featured as a case study in which I positioned her as a mathematician, slowly losing her interest in the subject. She features again in a conference presentation where I noticed her marginalisation in mathematics (Darragh 2015a), and she features in another article I am currently drafting. It is clear I recognised a richness of data in Emily. She was one of the more articulate of the participants involved in the study, but more importantly I was able to interpret her statements in ways that furthered my own agendas (to be published, to attend particular conferences) and desires to position myself as a particular kind of researcher. Here too we can see an effect of neoliberalism on the researcher, that is, the "satisfactions" and "rewards" of performativity (Ball 2013, p. 140).

Finally, I wish to mention that I also recognised Emily on a personal level. I liked her. She had a warm and open personality and was actively engaged in the research process. I could relate to the passion she expressed for mathematics in the first interview and I genuinely wished her success in the future. Yet at the same

---

[7]In the article referenced here, I call her Estelle. Emily and Estelle are both pseudonyms for the same person. This was not an intentional confusion!

time, in recognising Emily as Samoan girl I drew upon all the educational and statistical discourses which essentialise Samoan or Pasifika and gendered experiences. Furthermore, I recognised Emily primarily as a mathematics learner and I foregrounded this identity through the nature of my research. Ultimately I saw Emily as a research participant and saw her for her use value for me and my work.

4   Implications for Theory and Research Practice

In this paper I argue that identity research is a useful concept for social and political research in mathematics education, provided it is defined in a manner which incorporates its production within wider discourses. The performance of researcher is a powerful act and it works power not just over the subject, through identification practices, but also through the reconstitution of dominant (or subordinate) discourses. As Valero (2004) discusses, "we are implicated in constructing part of the practices of mathematics education in educational institutions when we act in those spaces as researchers" (p. 19), or in other words: "researchers are always already involved in what they critique, and hence, they produce knowledge and norms" (Llewellyn 2016, p. 303). This is not necessarily negative—the strengthening of some discourses and the disruption of others may be intentional and result in raising awareness within education of the marginalised experiences of particular groups of people. On the other hand, we run the risk of strengthening stereotypes. Was Emily's gender and ethnicity pertinent to my investigation? And, if so, was her social class, 'vulnerability' status, language abilities, race (as distinct from ethnicity), sexuality, or the fact she was able-bodied equally pertinent? The implication here is that we need to take care to consider the purposes and effects of our identifications, whether they be innocent seeming background descriptions or an intrinsic part of the research argument.

In deconstructing my identification of Emily my intent was not only to illustrate how researcher positioning is important, in concurrence with the literature mentioned earlier, but also to demonstrate that positioning and researcher identities are not static. The researcher is performing a temporal and multiple identity and this too needs to be understood as produced in discourse and generating of discourse. As researcher I am a performatively constituted subject, drawing from, constructing, and disrupting discourses. I do identity work with each spoken and written presentation and these identity performances are in part constructed by the stage, theatre, and audiences located in time and space. Finally, my identity performance in any moment also incorporates emotive aspects. In order to reveal fears/desires, purpose, constitutive discourses and aspects of identity, perhaps a final question to pose of the researcher is: 'What do I *enjoy* in this recognition of the participant in my research?' The responses are likely to be illuminative.

Acknowledgements Funding from: PIA-CONICYT Basal Funds for Centers of Excellence Project FB0003, and Project Fondecyt #3160469 are gratefully acknowledged.

References

Apple, M. W. (2000). Between neoliberalism and neoconservatism: Education and conservatism in a global context. In N. C. Burbules & C. A. Torres (Eds.), *Globalisation and education: Critical perspectives* (pp. 57–77). New York, NY: Routledge.

Ávalos-Salamé, D., & Thomas Ponce, E. (2007). *Medición de la vulnerabilidad social: Un análisis de los alumnos de Infocap*. Retrieved from http://www.infocap.cl/web/pdf/Informe_Final_Medicion_Vulnerabilidad_Social.pdf.

Ball, S. J. (2013). *Foucault, power, and education*. New York, NY: Taylor & Francis Inc.

Boaler, J., Wiliam, D., & Brown, M. (2000a). Students' experiences of ability grouping—Disaffection, polarisation and the construction of failure. *British Educational Research Journal, 26*(5), 631–648. https://doi.org/10.1080/01411920020007832.

Boaler, J., Wiliam, D., & Zevenbergen, R. (2000). The construction of identity in secondary mathematics education. In *International Mathematics Education and Society Conference*. Montechoro, Portugal. Retrieved from http://eprints.ioe.ac.uk/1142/1/Boalertheconstructionofidentity.pdf.

Boylan, M., & Povey, H. (2014). Ability thinking. In D. Leslie & H. Mendick (Eds.), *Debates in mathematics education* (pp. 7–16). London, UK: Routledge.

Braathe, H. J., & Solomon, Y. (2015a). Choosing mathematics: The narrative of the self as a site of agency. *Educational Studies in Mathematics, 89,* 151–166. https://doi.org/10.1007/s10649-014-9585-8.

Braathe, H. J., & Solomon, Y. (2015b). Choosing mathematics: The narrative of the self as a site of agency. *Educational Studies in Mathematics, 89,* 151–166.

Brown, T., Solomon, Y., & Williams, J. (2016). Theory in and for mathematics education: In pursuit of a critical agenda. *Educational Studies in Mathematics, 92,* 287–297. https://doi.org/10.1007/s10649-016-9706-7.

Butler, J. (1988). Performative acts and gender constitution: an essay in phenomenology and feminist theory. *Theatre Journal, 40*(4), 519–531. Retrieved from http://www.jstor.org/stable/3207893.

Butler, J. (1993). *Bodies that matter*. Oxon, United Kingdom: Routledge.

Butler, J. (1997). *The psychic life of power*. California: Stanford University Press.

Chassapis, D. (2017). "Numbers have the power" or the key role of numerical discourse in establishing a regime of truth about crisis in Greece. In *Mathematics education and life at times of crisis* (pp. 45–55).

Chronaki, A. (2011a). Disrupting "development" as the quality/equity discourse: Cyborgs and subalterns in school technoscience. In B. Atweh, M. Graven, W. Secada, & P. Valero (Eds.), *Mapping equity and quality in mathematics education* (pp. 3–19). Dordrecht, The Netherlands: Springer Science & Business Media.

Chronaki, A. (2011b). "Troubling" essentialist identities: performative mathematics and the politics of possibility. In M. Kontopodis, C. Wulf, & B. Fichtner (Eds.), *Children, development and education* (Vol. 3, pp. 207–226). Springer, The Netherlands. https://doi.org/10.1007/978-94-007-0243-1_13.

D'Ambrosio, B., Frankenstein, M., Gutierrez, R., Kastberg, S., Martin, D. B., Moschkovich, J., et al. (2013). Positioning oneself in mathematics education research. *Journal for Research in Mathematics Education, 44*(1), 11–22.

Damarin, S. K., & Erchick, D. (2010). Toward clarifying the meanings of gender in mathematics education research. *Journal for Research in Mathematics Education, 41*(4), 310–323.

Darragh, L. (2014). *Raising the curtain on mathematics identity: The drama of transition to secondary school*. University of Auckland.

Darragh, L. (2015a). Recognising gender in mathematics identity performances—Playing the fool? In S. Mukhopadhyay & B. Greer (Eds.), *Proceedings of the Eighth International Mathematics Education and Society Conference* (pp. 441–454). Portland, Oregon: MES8.

Darragh, L. (2015b). Recognising "good at mathematics": Using a performative lens for identity. *Mathematics Education Research Journal, 27*(1), 83–102.

Darragh, L. (2016). Identity research in mathematics education. *Educational Studies in Mathematics, 93*, 19–33. https://doi.org/10.1007/s10649-016-9696-5.

Davison, I. (2013). Gap widens between NZ students. *The New Zealand Herald*. Retrieved from http://www.nzherald.co.nz/nz/news/article.cfm?c_id=1&objectid=11167148.

de Freitas, E. (2008). Mathematics and its other: (Dis)locating the feminine. *Gender and Education 2, 20*(3), 281–290.

Gutiérrez, R. (2013). The sociopolitical turn in mathematics education. *Journal for Research in Mathematics Education, 44*(1), 37. Retrieved from http://www.jstor.org/stable/10.5951/jresematheduc.44.1.0037.

Gutiérrez, R., & Dixon-Román, E. (2011). Beyond gap gazing: How can thinking about education comprehensively help us (re)envision mathematics education? In B. Atweh, M. Graven, W. Secada, & P. Valero (Eds.), *Mapping equity and quality in mathematics education* (pp. 21–34). Dordrecht, The Netherlands: Springer.

Harvey, D. (2005). *A brief history of neoliberalism*. Oxford, UK: Oxford University Press.

Hossain, S., Mendick, H., & Adler, J. (2013). Troubling "understanding mathematics in-depth": Its role in the identity work of student-teachers in England. *Educational Studies in Mathematics, 84*(1), 35–48. https://doi.org/10.1007/s10649-013-9474-6.

Jurdak, M., Vithal, R., de Freitas, E., Gates, P., & Kollosche, D. (2016). *Social and Political dimensions of mathematics education: Current thinking*. Switzerland: Springer Open.

Kanes, C., Morgan, C., & Tsatsaroni, A. (2014). The PISA mathematics regime: Knowledge structures and practices of the self. *Educational Studies in Mathematics, 87*, 145–165. https://doi.org/10.1007/s10649-014-9542-6.

Lerman, S. (2009). Pedagogy, discourse, and identity. In L. Black, H. Mendick, & Y. Solomon (Eds.), *Mathematical relationships in education*. New York: Routledge.

Llewellyn, A. (2008). "Maths with Sam and Alex": A discussion of choice, control and confidence. In J. F. Matos, P. Valero, & K. Yasukawa (Eds.), *Proceedings of the Fifth International Mathematics Education and Society Conference* (pp. 362–375). Lisbon: Department of Education, Learning and Philosophy, Aalborg University.

Llewellyn, A. (2009). "Gender games": A post-structural exploration of the prospective teacher, mathematics and identity. *Journal of Mathematics Teacher Education, 12*(6), 411–426. https://doi.org/10.1007/s10857-009-9109-0.

Llewellyn, A. (2016). Problematising the pursuit of progress in mathematics education. *Educational Studies in Mathematics, 92*(3), 299–314. https://doi.org/10.1007/s10649-015-9645-8.

Lundin, S. (2012). Hating school, loving mathematics: On the ideological function of critique and reform in mathematics education. *Educational Studies in Mathematics, 2*(80), 73–85.

Martin, D. B. (2000). *Mathematics success and failure among African-American youth*. New Jersey: Lawrence Erlbaum Associates.

Martin, D. B. (2009). Researching race in mathematics education. *Teachers College Record, 111*(2), 295–338.

Martin, D. B., Gholson, Maisie, & Leonard, L. (2010). Mathematics as gatekeeper: Power and privilege in the production of knowledge. *Journal of Urban Mathematics Education, 3*(2), 12–24.

Mead, G. H. (1913). The social self. *The Journal of Philosophy, Psychology and Scientific Methods, 10*(14), 374–380. https://doi.org/10.2307/2012910.

Mendick, H. (2005). Mathematical stories: Why do more boys than girls choose to study mathematics at AS-level in England? *British Journal of Sociology of Education, 26*(2), 235–251. https://doi.org/10.1080/0142569042000294192.

Mendick, H. (2017). Mathematical futures: Discourses of mathematics in fictions of the post-2008 financial crisis. In A. Chronaki (Ed.), *Mathematics Education and Life at Times of Crisis: Proceedings of the Ninth International Mathematics Education and Society Conference* (pp. 74–89). Volos, Greece: MES9.

Mendick, H., Epstein, D., & Moreau, M.-P. (2008). Mathematical images and identities: Entertainment, education, social justice. *Research in Mathematics Education, 10*(1), 101–102. https://doi.org/10.1080/14794800801916978.

Nasir, N. S. (2002). Identity, goals, and learning: Mathematics in cultural practice. *Mathematical Thinking and Learning, 4*(2&3), 213–247.

Pais, A., & Valero, P. (2012). Researching research: Mathematics education in the political. *Educational Studies in Mathematics, 80*, 9–24. https://doi.org/10.1007/s10649-012-9399-5.

Pais, A., & Valero, P. (2014). Whither social theory. *Educational Studies in Mathematics, 87*(2), 241–248. https://doi.org/10.1007/s10649-014-9573-z.

Planas, N., & Valero, P. (2016). Tracing the socio-cultural-political axis in understanding mathematics education. In *The second handbook of research on the psychology of mathematics education* (pp. 447–479).

Robertson, S. (2012). Placing teachers in global governance agendas. *Comparative Education Review, 56*(4), 584–607. Retrieved from http://www.jstor.org/stable/10.1086/667414.

Rubie-Davies, C. M. (2006). Teacher expectations and student self-perceptions: Exploring relationships. *Psychology in the Schools, 43*(5), 537–552.

Solomon, Y., Lawson, D., & Croft, T. (2011). Dealing with "fragile identities": Resistance and refiguring in women mathematics students. *Gender and Education, iFirst Art*, 1–19. https://doi.org/10.1080/09540253.2010.512270.

Solomon, Y., Radovic, D., & Black, L. (2015). "I can actually be very feminine here": Contradiction and hybridity in becoming a female mathematician. *Educational Studies in Mathematics*. Published https://doi.org/10.1007/s10649-015-9649-4.

Stinson, D. W. (2008). Negotiating sociocultural discourses: The counter-storytelling of academically (and mathematically) successful African American male students. *American Educational Research Journal, 45*(4), 975–1010. Retrieved from http://ezproxy.auckland.ac.nz/login?url=http://search.proquest.com/docview/200446519?accountid=8424.

Stinson, D. W. (2013). Negotiating the "White Male Math Myth": African American male students and success in school mathematics. *Journal for Research in Mathematics Education, 44*(1), 69. Retrieved from http://ezproxy.auckland.ac.nz/login?url=http://search.proquest.com/docview/1284416138?accountid=8424.

Straehler-Pohl, H., Gellert, U., Fernandez, S., & Figueiras, L. (2014). School mathematics registers in a context of low academic expectations. *Educational Studies in Mathematics, 85*(2), 175–199.

Swanson, D., & Black, L. (2017). Integrating critical theory and practice in mathematics education. In *CERME10*. Dublin.

Turner, H., & Rubie-Davies, C. M. (2015). Teacher expectations, ethnicity and the achievement gap. *New Zealand Journal of Educational Studies*. https://doi.org/10.1007/s40841-015-0004-1.

Vale, C., Atweh, B., Averill, R., & Skourdoumbis, A. (2016). Equity, social justice and ethics in mathematics education. In K. Makar, S. Dole, J. Visnovska, M. Goos, A. Bennison, & K. Fry (Eds.), *Research in mathematics education in Australasia 2012–2015* (pp. 97–118). Singapore: Springer Nature. https://doi.org/10.1007/978-981-10-1419-2.

Valenzuela, J. P., Bellei, C., & de los Ríos, D. (2014). Socioeconomic school segregation in a marketoriented educational system. The case of Chile. *Journal of Education Policy, 29*(2), 217–241. https://doi.org/10.1080/02680939.2013.806995.

Valero, P. (2004). Socio political perspectives on mathematics education. In P. Valero & R. Zevenbergen (Eds.), *Researching the socio political dimensions of mathematics education*. Boston: Kluwer Academic Publishers.

Valero, P. (2016). Mathematics for all, economic growth, and the making of the citizen-worker. In T. Popkewitz, J. Diaz, & C. Kirchgasler (Eds.), *Political sociology and transnational educational studies: The styles of reason governing teaching, curriculum and teacher education*. London, UK: Routledge.

Walkerdine, V. (1990). *Schoolgirl fictions*. London, UK: Verso.

Walkerdine, V. (1994). Reasoning in a post-modern age. In P. Ernest (Ed.), *Mathematics, education and philosophy: An international perspective*. London: The Falmer Press.

Walkerdine, V. (1998). *Counting girls out: Girls and mathematics* (new ed.). London: Falmer Press.

Walshaw, M. (2001). A Foucauldian gaze on gender research: What do you do when confronted with the tunnel at the end of the light? *Journal for Research in Mathematics Education, 32*(5), 471–492.

Walshaw, M. (2010). The researcher's self in research: Confronting issues about knowing and understanding others. In L. Sparrow, B. Kissane, & C. Hurst (Eds.), *Shaping the Future of Mathematics Education: Proceedings of the 33rd annual Conference of the Mathematics Education Research Group of Australasia* (pp. 587–593). Fremantle, Australia: MERGA.

Walshaw, M. (2013). Post-structuralism and ethical practical action: Issues of identity and power. *Journal for Research in Mathematics Education, 44*(1), 100–118. https://doi.org/10.5951/jresemmatheduc.44.1.0100.

Webb, A., & Radcliffe, S. (2013). Mapuche demands during educational reform, the Penguin revolution and the Chilean winter of discontent. *Studies in Ethnicity and Nationalism, 13*(3), 319–341.

William, D., Bartholomew, H., & Reay, D. (2004). Assessment, learning and identity. In P. Valero & R. Zevenbergen (Eds.), *Researching the socio political dimensions of mathematics education*. Boston: Kluwer Academic Publishers.

Youdell, D. (2006a). Diversity, inequality, and a post-structural politics for education. *Discourse: Studies in the Cultural Politics of Education, 27*(1), 33–42. https://doi.org/10.1080/01596300500510252.

Youdell, D. (2006b). Subjectiviation and performative politics—Butler thinking Althusser and Foucault: Intelligibility, agency, and the raced-nationed-religioned subjects of education. *British Journal of Sociology of Education, 27*(4), 511–528.

Zeichner, K. (2010). Competition, economic rationalization, increased surveillance, and attacks on diversity: Neo-liberalism and the transformation of teacher education in the U.S. *Teaching and Teacher Education, 26*, 1544–1552. https://doi.org/10.1016/j.tate.2010.06.004.

Zevenbergen, R. (2003). Ability grouping in mathematics classrooms: A Bourdieuian analysis. *For the Learning of Mathematics, 23*(3), 5–10.

Truth, Power and Capitalist Accumulation in Mathematics Education

Alexandre Pais

Abstract In this chapter I raise a set of questions intended to make us reflect on our work as researchers, namely in the way we propagate and naturalise common assumptions or truths about mathematics education, as well as the mechanisms of power that makes it difficult for us to see beyond these well-accepted truths. I suggest that some of the forces that impact upon and restrict socially just outcomes for mathematics education are not just "external", that is, originated outside the mathematics education community, but also, and perhaps more importantly for us, from the way research itself addresses the teaching and learning of mathematics in schools. Instead of positing ourselves as the beautiful souls of mathematics education, my invitation is for us to posit ourselves as part of the problem, and be willing to address some of our ideological assumptions before relegating to the social and political world the causes of our discontentment. For this purpose, I will rely on Foucault's and Lacan's works on the notion of truth, as a way to explore the role that contemporary mathematics education plays within capitalism.

Keywords Truth · Power · Capitalist accumulation · Foucault
Lacan

1 Introduction

This chapter originates from an invitation made by the organisers of the ICME-13 Topic Study Group on the *Social and political dimensions of mathematics education*, to address issues of power and truth in mathematics education—timely questions that involve all of us who work in mathematics education. In my work, I have been arguing for the importance of researchers to address not only questions that concern the learning and teaching of mathematics in schools, but also the role of research itself in the creation and maintenance of much of the predicaments that

A. Pais (✉)
Manchester Metropolitan University, Manchester, UK
e-mail: a.pais@mmu.ac.uk

© Springer International Publishing AG 2018
M. Jurdak and R. Vithal (eds.), *Sociopolitical Dimensions of Mathematics Education*, ICME-13 Monographs, https://doi.org/10.1007/978-3-319-72610-6_6

characterise the field. It is my contention that without a critical reflection that posits research as part of the ongoing problem of failure in school mathematics, researchers lack a meaningful mapping of the key challenges for mathematics education. As such, this chapter raises a set of questions intended to make us reflect on our work as researchers, namely in the way we propagate and naturalise common assumptions or truths about mathematics education, as well as the mechanisms of power that makes it difficult for us to see beyond these well-accepted truths.

In what follows, I will briefly present and discuss what I consider to be some of the keystone truths of mathematics education, and outline the mechanisms of power that allow for the sedimentation of these truths. Throughout the text, I will address some of the questions raised in the Topic Study Group call. I will propose that some of the forces which impact upon and restrict socially just outcomes for mathematics education are not just "external", that is, from outside the mathematics education community, but also, and perhaps more importantly for us, from the way research itself addresses the teaching and learning of mathematics in schools. Instead of positing ourselves as the *beautiful souls* (Pais 2017) of mathematics education, my invitation is for us to posit ourselves as part of the problem, and be willing to address some of our ideological assumptions before relegating to the social and political world the causes of our discontentment.

2 Truths in Mathematics Education

What are the truths en vogue today in the mathematics education community? Although being a large and highly diversified field of research (for ICME-13 alone there were over 50 Topic Study Groups in different areas), there are common shared assumptions that most of the people working with mathematics education assume. These can be called the truths of the field, in the sense that they provide a common platform on which all agree, notwithstanding the array of different practical, methodological or theoretical approaches. These truths often remain un-theorised. They are rather taken for granted as "evident" or posited as an ideal to be achieved. In this section I will discuss five of these truths, informed by some of my previous work.

2.1 School Mathematics Should Be Enjoyable

Mathematics is at once a corner stone of modernity and a headache. Its paramount presence in modern achievements, its role in providing a language for science and technology, and its importance for economic development, contrasts with the renowned lack of knowledge, if not aversion, that a significant part of the world population holds for this subject. School mathematics is often portrayed as a difficult and unpleasant subject not only by students, but also by teachers who see themselves in the situation of having to teach something they are not quite comfortable with.[1] Against this reality, researchers struggle to turn mathematics into the object of students' (and teachers') desire. In our days, it is not enough for a student to do well in mathematics, passing the exam and moving on with life. Students also have to enjoy or even to love (Boaler 2010) mathematics. As expressed by Cobb (2007), students should learn mathematics in order to develop an "empathy for a sense of affiliation with mathematics together with the desire and capability to learn more about mathematics when the opportunity arises" (p. 9). The purpose seems to be not only to guarantee that students learn mathematics in a meaningful way, but also in an *enjoyable* way. Curricular reforms around the world have been trying to change the negative image of mathematics by presenting it as something interesting, related to students' reality, and enjoyable, as something that could be fun.[2]

Although alluring in prospect, a Foucauldian approach to school as an institution concerned with normalisation, and the curriculum as a system of reason, allows for a critique of the idea that mathematics should be enjoyable. Curricular reforms that posit mathematics as enjoyable and fun show how education concerns not only knowledge and competences, but also the innermost feelings of the students (empathy, desire, love, delight, etc.)—thus constituting what Foucault (1997) calls a *technology of the self* (Foucault 1997) aimed at fabricating the kind of subjects susceptible to being governed (Popkewitz 2004). According to Fendler (1998), the purpose of current educational systems is to govern the soul. Teachers have not only the responsibility to govern the moral, but also the feelings, the desires, and anxieties, in order to produce the wanted citizen: "becoming educated, in the current sense, consists of teaching the soul—including fears, attitudes, will, and desire" (p. 28). Although presented as a liberating and emancipatory experience, the appeal to enjoy mathematics conceals a deeper attempt to control not only people's knowledge, but also, and perhaps more importantly, their feelings.

2.2 People Use Mathematics in Their Daily Lives

Common sense says that people do not use mathematics in their daily lives. Research often confirms this unimportance of mathematics for mundane activities (e.g., Brenner 1998; Jurdak 2006; Williams and Wake 2007). However, instead of questioning the presupposition that people need mathematics for their mundane or professional activities, research takes to itself the task of improving the utility of

---

[1]This is particularly the case with primary teachers who, besides mathematics, have to teach all the other subjects.

[2]See, for instance, the United Kingdom, Portugal and Sweden curriculums, where enjoyment is posited as one of the main goals for the teaching and learning of mathematics: "the subject aims at pupils experiencing delight in developing their mathematical creativity, and the ability to solve problems, as well as experience something of the beauty and logic of mathematics" (Utbildningsdepartementet 2000).

mathematics (Lundin 2012). This is done by means of developing deeper analysis and positive experiences whereby students actually transfer mathematics from and into school: people do not use mathematics, but (because they should) we need to continue developing efforts to change this situation.

Elsewhere I analyse in depth the ideology perpetrating the assumption that mathematics is important as use-value (Pais 2013). Here, I take Murad Jurdak's research as an example of the functioning of this ideology. After concluding that "the activity of situated problem solving in the school context seems to be fundamentally different from decision-making in the real world because of the difference of the activity systems that govern them" (Jurdak 2006, p. 296), and that students "define their own problems, operate under different constraints, and mathematics, if used at all, plays a minor role in their decision making" (p. 296), Jurdak nevertheless, insists on the importance of confronting students with real-life situations: "simulations of such authentic real life situations as embedded in situated problem solving may provide a plausible option to develop appreciation of the role, power, and limitations of mathematics in real world decision-making" (p. 296). He adds, "though *quite different* in real life from that in school, the process of mathematization is *essentially the same* and having experience in it in a school context may impact on mathematization in real life" (p. 297, my emphasis). When confronted with the difficulties in transfer, Jurdak proceeds by eliminating the obstacles, so that the higher goal of making mathematics useful for people's lives can be kept. Instead of assuming the impossibility of transfer (Evans 1999; Gerofsky 2010), the researcher ends up creating an ideology whose purpose is precisely to disavow such impossibility. It is impossible to find support in the research reported in Jurdak's text for such statements. The belief that the exploration of real-life situations in school will impact on the way in which people use mathematics in real life is based on a leap of "faith" (Lundin 2012), thus ideology at its purest. The emergent question is thus: why do researchers continue to argue for the importance of mathematics as use-value, notwithstanding all the evidence that people do not use school mathematics in their daily lives?

2.3   Mathematics for All

Currently, the ideal of a "mathematics for all" is systematically (and uncritically) foregrounded as the ultimate horizon guiding our engagement in the field (Pais 2012, 2014). A slogan such as "mathematics for all", functions as a *master-signifier* (Žižek 2012), a banner upon which we all agree, uniting the field, thus offering a space whereby different perspectives, theories and methodologies, can "work together". "Mathematics for all" can be seen as a fantasy formation, whose purpose is not (only) to make sense of the world in a wholly way, but precisely to conceal the impossibility of making sense of it. Although we know that mathematics is not for all, that it serves other purposes than the ones related with knowledge and competences, and that many students find it meaningless or even traumatic, we rely on the illusion that mathematics can indeed be for all, that it can be an adventure into knowledge, and a pleasurable and useful subject for students. The shocking evidence that school mathematics is nothing of this does not inhibit from partaking in the illusion that it can indeed be so. As a result, instead of asking why is it not so, we keep researching how can it be so.

In a recent conversation with a colleague around these issues, he claimed that although we know very well that mathematics is not for all, we should refrain ourselves from saying it out load. Admitting that mathematics is not for all will potentiality diminishing its importance in schooling (who says "geography for all"?), with direct consequences for our work as researchers. It is because mathematics plays such a relevant role in society and schooling that we, as a research community, enjoy privileged funding and working opportunities. As I explore elsewhere (Pais 2017), what this discourse renders evident, however, is how research is about nothing but itself. It seems that research is not about improving school mathematics, but about using the miserable state of school mathematics to give researchers conditions to develop their work.

2.4 Researching Success

Notwithstanding all the evidence that mathematics is not for all, this ideal is posited as an achievable goal, and emphasis is given to the exploration of successful experiments, where students seem to learn meaningful mathematics for their lives. To develop and broadcast successful experiences seems to be the aim of research (e.g. Gutiérrez 2010; Sriraman and English 2010).

What can classroom examples say to us about the ideal of "mathematics for all"? As any teacher knows, in a class of thirty students there will always be some (often many) who fail. The crude reality tells us that the ideal is at least an illusion (when not a straightforward bait). In order to enable success, however, researchers set and organise classroom data in a way that can corroborate a priori assumptions. As Paola and I explore (Pais and Valero 2012), in Luis Radford's theory of cultural objectification, for instance, the examples used to support the theory (2006, 2008) are all reports of successful experiences, whereby pupils always acquire (objectify) the mathematical content demanded by the teacher. The research environment is set in a way as to avoid friction and allow a meaningful mathematics learning to occur, and the classroom examples are chosen to fit the theory. Radford's theory of objectification drifts at the very moment we try to imagine it applied in low-streaming schools in Germany (Straehler-Pohl and Pais 2014), schools in post-apartheid South-Africa (Skovsmose and Valero 2008), ghetto schools in the US (Gutstein 2003), or even a public European school struggling with imposed forms of mathematics that do not match the safeness and aseptic schooling characteristic of Radford's research settings (Brown 2011). In these settings, very seldom do students "unite" (Radford 2006, p. 54) with the culture of mathematics in the way envisage by Radford's theory. Contrary, what often occurs is precisely a refusal to identify with the mathematical successful learner envisaged by the curriculum (Pais 2016).

For research to break with this "epistemological obstacle" (Bachelard 2002) it needs to seriously take its object of study—the teaching and learning of mathematics—as "it is" instead of how it "ought to be" (Pais and Valero 2014). Moreover, it has to posit its object as the very arena in which research assumptions are tested. This implies moving from questioning "what can a school do if it wants to engage all of its students actively and productively in relevant mathematics learning?" (Clements 2013, p. ix), to questioning why schools cannot systematically engage all of its students actively and productively in relevant mathematics learning, notwithstanding the declared will of all involved. In other words, instead of seeing research as a mean to change practice, perhaps researchers should take practice itself—as it happens in most schools, outside the fixed environments designed by researchers—as a mean to change research theories, methodologies and approaches.

2.5 Research Improves Practice

The fifth and final truth concerns the gap between research and practice (Sriraman and English 2010). In the introductory chapter of the Third International Handbook of Mathematics Education, Clements (2013) poses a crucial question for all of those involved in mathematics education research:

> Why has there not been a marked improvement, given the large amount of mathematics education research conducted around the world, and over a very long period of time, with respect to such fundamentally important curriculum matters? (p. x, xi).

Given that mathematics education as a field of research is not only oriented to describing and analysing practice, but (and perhaps more importantly) to prescribe or at least identify good practice (Jablonka et al. 2013, p. 47), this situation is worrisome.

As I explore elsewhere (Pais and Valero 2012), the discrepancy between the sophistication of research and the lack of change in school mathematics is often displaced from research and posited on the way governments, schools and teachers fail to "acquire" and implement the knowledge originating from the academia. In research, everything goes well; we know the best methods, theories and strategies. The problems of implementation rest in the school settings. Lundin (2012) has recently discussed the fallacy of this line of argumentation. What he calls *the standard critique of mathematics education* consists of describing the current state of affairs of school mathematics as suffering from a variety of malfunctions, and the role of mathematics education research to fix them. The problem with this argumentation is that it eschews research from a critical analysis of its own role in the creation of the very same gap that it so eagerly strives to close. As argued by Klette

(2004), the problem of change in mathematics education reforms is not just a problem of "application" but may very well be an embedded part of research itself. She argues that the "denial of change" (p. 3) is being constructed from the beginning, in the theoretical, methodological and conceptual ways in which research is done.

3 Power in Mathematics Education

If we undertake a Foucauldian reading of truth and power, then the question to be asked is: what are the power relations at work in the production of such truths in mathematics education? Foucault (1979) suggests that the production and maintenance of a truth, instead of deriving from some universal knowledge about the world, is rather the result of particular individual interests. In a way, for Foucault, we tend to adopt the truths that are more convenient for the achievement of our own goals. As researchers, we cannot be blind to the fact that there are obvious benefits from the belief that mathematics is precious knowledge, a keystone of modern society, and an inescapable tool for citizenship. On the other hand, by positing mathematics for all as a goal to be achieved, and by asserting the importance of research in this process (against the malaises of practice) we set the ideological frame wherein we can continue to work, receiving our salaries, progressing in our careers, participating in conferences, travelling, enjoying ourselves. Such are the relations of power in which we are all involved, and which produce the truths that we take for granted when thinking about mathematics education.

As noticed by Foucault (1979), power is only exercised between free subjects, who might not recognise themselves as actors of power. Moreover, power is not a substance that can be deposited in subjects (the non-Foucauldian notion of *empower*) or kept by some sovereign figure (the typical case here being the monarch). The main objective when analysing power relations is not so much to decipher how power is present in "such or such" institution, or group, or elite, or class but rather how all the individuals "freely" participate in a certain technique or exercise of power. An analysis of the power relations in mathematics education will thus refrain from framing the problem in terms of a struggle between those who have power (the usual suspects: governments, bureaucrats, regulatory agencies) and those who have not (researchers, teachers, students). Instead, Foucault invites us to posit ourselves as part of the problem, as free subjects that participate in power relations within a certain structural arrangement.

This approach to power however contrasts with the way in which power is usually perceived in mathematics education. As analysed by Skovsmose and Valero (2008), mathematics education is seen as something through which people can be *empowered* (e.g. NCTM 2016). Mathematics gives power to people, whether through the intrinsic characteristics of mathematics itself (logical thinking, abstraction); by providing students with psychological meaningful experiences (solving problems, metacognition); by enhancing the relation between cultural background and foreground therefore allowing students to learn 'in context' (connection between every day practices and school mathematics; providing opportunities to envision a desirable range of future possibilities); or by exploring situations of 'mathematics in action' which makes visible the way mathematics formats reality (exploring real mathematical models in a critical way). Such an approach to power, as a substance that "empowers", is at odds with Foucault's analysis, where power is instead to be perceived as circulating through a microphysics of practice. The very attempt made by the mathematics education community to empower people through mathematics disavows a more structural understanding of how this same effort is already part of a power relation and a regime of truth that posits mathematics as an important knowledge and competence to master the world. What researchers miss to recognise (or accept) is the way in which mathematics empowers people not so much because it provides some kind of knowledge or competence to them, but because it gives people a *value*. It allows students to accumulate credit in the school system that will allow them to continue studying and later to achieve a comfortable place in the economic and social order. Mathematics empowers people because it is posited as an economically valuable resource.

Elsewhere I have shown how the discourse around the importance of mathematics as knowledge and competences constitutes an ideology set on effacing the role which school mathematics plays in political economy (Pais 2013, 2015). However, to assume that school mathematics is more about credit than about mathematics (Baldino and Cabral 2013; Pais 2012) implies questioning the entire discourse sustaining mathematics education research, thus jeopardizing the central role mathematics has in education, with all the consequences this will have for our work. The crude reality that for many people around the world mathematics is no more than a meaningless school subject that they need to pass in order to go on with life that it is not for all, and that research grows irrespectively of what is happening in schools, must remain either silent or conceived as an obstacle likely to be solved through better research and teaching practices. The challenge is thus to posit these "malfunctions" as the concrete truths of today's mathematics education.

4   Truth in Mathematics Education

Such a move encompasses a conceptualization of truth different from the Foucauldian one. Foucault was interested in deciphering the mechanisms (the regimes) by which a statement becomes perceived as being "true", that is, accepted as natural and beyond questioning. The five truths that I have elicited are examples of such approach to truth in mathematics education. In this vein, truth is something that is all too visible, it is everywhere. The challenge is to analyse how what appears as truth is indeed the result of a historical and discursive process that, far from being natural, is born out of regimes of interests and power relations.

French psychoanalyst Jacques Lacan introduces a somehow different conceptualisation of truth. For Lacan truth manifests itself through what fails (Lacan 1990). If we consider an analysand freely speaking with her analyst within the context of an analysis, truth is not to be confused with the narratives that she tells about herself (episodes from her childhood, the story of her marriage, parenthood, professional achievements and deadlocks, etc.). Instead, the truth of her discourse only emerges surreptitiously through slips of tongue, puns, silences, or any other impasses in her discourse. These "malfunctions" that break the flow of the ego's narrative are the truth about the subject. We cannot directly understand them (contrary to the truth-narrative that the patient tells, which is effortlessly understandable), rather they signal an inconsistency that stands for truth as such.

What can we infer if we apply this notion of truth to mathematics education research? I suggest that the truth of mathematics education is not to be found in the official narrative ("mathematics for all", the importance of mathematics for mundane activities, the idea that research improves practice, etc.) but precisely in the hindrances that disturb or do not fit into this narrative (the student who refuses to learn, the persistence of failure in school mathematics, the evidence that people do not use mathematics in their daily lives, etc.). All these obstacles tend to be foreclosed by research, by creating the narrative that mathematics is for all, by organising classroom arrangements where students apparently learn important mathematics for their lives, or by continuously assuming that more and better research is needed to improve school mathematics. To let the truth speak means to pay attention to what is failing in mathematics education.

Against this background, one of the first implications for research is to study what *fails*. As I explore elsewhere (Pais 2014), researchers tend to focus on the exploration of successful experiences. It will not be easy for the reader to find a study that takes failure in itself and uses it to shed light on the contradictions of the whole system.[3] Research is animated by a sense of "positivity", and values situations where, notwithstanding all the difficulties, a breakthrough was possible (Gutiérrez 2010; Presmeg and Radford 2008; Sriraman and English 2010). As posited by Gutiérrez, "it is important to highlight the features of practice that coincide with certain kinds of students engaging/succeeding in school mathematics (and this form is much more productive than focusing on failure and/or disengagement)" (2010, p. 52). Though this approach may be convenient, it makes impossible a broader critique of the equity model in which current schooling is based. Moreover, it provides the ideological frame against which researchers can continue doing their work without questioning the economically rooted reasons of failure.

---

[3] An exception are the works or Roberto Baldino and Tânia Cabral.

5   Truth, Accumulation and Capitalism

At stake here is what Lacan (2007) calls the logic of *accumulation* that characterizes modern science under the auspices of capitalist economy. Within capitalism, any measure has to produce surplus-value, otherwise it is discarded as obsolete, against the rules of the market, and the like.[4] And the same with science. Any scientific result that threatens the homogeneity of science, its corpus of truth, results in a crisis. Modern science is built as an accumulative regime of knowledge, inasmuch as capitalist economy has at its core, the accumulation of capital. Any threat to this cycle of accumulation is seen as irrational, retrograde, and even impossible.[5]

As recently explored by Lacanian philosopher Samo Tomšič (2015), one way of describing capitalism is that it is life without negativity, that is, the efficiency and the logic of capitalism is supported by a fantasy/ideology of a subjectivity and a society without negativity (p. 7).[6] Capital presupposes a life without subjects, only individuals, acting according to the social demand, to the social place assigned to them. The capitalist system has difficulties to deal with people that somehow do not fit into what is expected them to be, from individual cases such as Julian Assange or Nadezhda Tolokonnikova, to great masses of refugees and migrants that stand for what Rancière (1995) called the *part of no part* of the current political order (and, as a no part, end up dying in the Mediterranean sea or exploited into slavery as is happening now at the core of the European Union). These singularities have the potential to point towards the inconsistencies of the entire system, thus allowing for a questioning of capitalism as the global structural arrangement in place today. However, as pointed out by Tomšič, "the subject's non-identity is perceived as secondary and as something that could be abolished simply by 'correcting' the structural relations that dim brought the subject into existence" (p. 65). Examples of these "corrections" are the EU-Turkey agreement to deal with the refugee crisis—where the European Union paid billions of euros for the Turkish government to keep refugees outside Europe—or the entire industry around charity and philanthropy (with the United Nations leading the way). Capitalism produces a "world-view", a reality that appears to function (albeit some correctable malfunctions), a reality without lack or negativity (Tomšič 2015, p. 96). In other words, and to recover our previous discussion, a reality that does not burden itself with truth.

The same logic of foreclosure of the subject and of truth is at work in modern science. Lacan attributes science's prodigious fecundity to the fact that it wants to

---

[4]Suffice to think about the hysterical reaction every time someone suggests an increase in social benefits, a reduction of the working hours, or a public investment in public healthcare and education.

[5]The history of modern science is rich in episodes that show how difficult it is for results that do not fit into a certain stablished worldview to be accepted—Copernicus' model of the universe is perhaps the most well-known example, but we can also mention the introduction of non-Euclidian geometries and non-standard analysis in mathematics, Darwin's theory of natural evolution in biology, Marx's works on political economy or Freud's studies of the human psyche.

[6]In this sense, capital is creative potential, a specific form of vitalism.

know nothing about truth as cause (Skriabine 2013, p. 52). Truth as a cause is not to be confused with truth as *adequatio rei intellectus*, correspondence of the thing to the mind—in which science has thrived. Instead, it signals the potential that science has to explore what fails, what cannot be immediately assimilated into a given "worldview", without a radical change of that worldview itself. Such a gesture goes against the accumulative spirit of science, where attempts to re-consider the direction of scientific development and knowledge production are seen as backward and negative. What is important is to keep the cycle of scientific production and accumulation non-interrupted. The challenge for science today is to mobilise the subversive dimension of modern science, its inherent *negative* core, its *passion for truth*. Not the Foucauldian "factual" truth, but truth as such in the gap of a certain knowledge.

As I have been showing throughout this chapter, mathematics education, as a science, is not immune to this drive towards accumulation, noticeable in the way it privileges the exploration of successful situations, the demand to produce implications for agents and institutions, the pressure to publish articles and books that contribute to the enlargement of the field. All this keeps the system running, thus disavowing a questioning of the entire purpose of the educational industry that mathematics education has become.

6 Final Remarks: Revisiting the Five Truths

The five truths previously discussed function in a way as to disavow the truth of school (mathematics). To use Althusser's distinction between ideology and science (Althusser 2008), the truths of the field keep mathematics education at the level of ideology. This is what elsewhere I called the *narcissism of mathematics education* (Pais 2017), where, by ignoring the concrete problems experienced by teachers and students in schools, mathematics education research only reinforces its own prejudices about school mathematics. My plea is for researchers to pay attention to the evidence coming from schools, and use it to confront their assumptions about the importance of both school mathematics and of research. A plea in all identical to the one made by Max Planck to his fellow physicians colleagues one hundred years ago when confronted with the inadequacies of classic mechanics to explain the results emerging from quantum experimentations. Instead of trying to fit what we observe into an already formed frame (one informed by the five truths), we need to take what we observe seriously and build research that fits these observations. Researchers need to be dragged from their complacent truths by the brute facts observed in their laboratories—schools. Examples of such brute facts were already explored in this chapter, and they include the endemic nature of failure in school mathematics, the fact that people do not use mathematics outside school, that research does not improve practice, and that mathematics is not for all. To take reality as it is, with all its tensions and contradictions—instead of organising experiments where everything is set up as to guarantee that a meaningful mathematics education will occur, what Skovsmose calls the "prototypical class-rooms" of research (2005)—is the only way for mathematics education to evolve as a science. This can be a painful process, as it implies questioning the very some truths that currently sustain our work as researchers.

In the guise of conclusion, I revisit the five truths previously listed and elaborate, for each one, concrete implications that many of us can start implementing next Monday morning.

6.1 Mathematics Should Be Enjoyable

My suggestion is for us to assume that mathematics is for many students a tasteless, meaningless and even traumatic school subject that they need to pass in order to carry on with their lives. Students might need to do school mathematics as part of their education, but they do not have to like it.

6.2 The Importance of Mathematics

I suggest placing the importance of mathematics not in terms of its inherent characteristics—problem solving, utility, beauty, cultural possibilities, etc.—but in terms of its attendant submissions to economic criteria and goals. In Pais (2013, 2014) I argue that by positing the importance of school mathematics in terms of *knowledge* and *competence*, research provides an ideological screen against the role school mathematics plays within capitalist schooling. My suggestion is to conceive the importance of mathematics not in terms of mathematics itself, but in terms of the place this subject occupies within a given structural arrangement, that is, in terms of the *value* that school mathematics has. As teachers, it implies being honest with our students and openly say that not only they do not have to "like" mathematics, its presence in the school curriculum has nothing to do with its use value, but with the value that they will get from passing the exam. Share our own contradictions with students and let them be aware of our own doubts as teachers and our criticism of schooling. Again, we might be obliged to perform certain tasks (like doing routine and stupid exercises to prepare students for an exam), but we do not have to like it.

6.3 Mathematics for All

We need to assume that failure is endemic to schooling. Instead of running after the hysterical societal demand of mathematical equity, developing increasingly refined stratagems to better teach and learn mathematics that only seem to function in the controlled reality of a research setting, perhaps we should acknowledge the crude reality that mathematics is not for all. Schools, however uncomfortable such awareness may be, are places of selection and teachers are agents of exclusion. These are the conditions of today's schooling, and research cannot afford dismissing them as being beyond its field of action. Publicly assuming that mathematics is not for all may not solve any problem, but at least does not mask it.

6.4 Researching Success

See Sect. 4. There is a need to research what fails not only in the practice of others but also in our own practice as researchers.

6.5 Research Improves Practice

If researchers know so well what needs to change in school mathematics, perhaps they should go to schools and work as teachers. It takes a lot of imagination to understand how researchers can change the problems of school mathematics. If teachers, being the ones who know the students, the school, their families, the community, cannot solve the problems, how can a researcher, who is not immersed in the school, do it? On the other hand, teachers will benefit from having more time to develop their own research, including researching their own practice. My suggestion is thus to equally distribute both teaching and research among teachers and researchers. Every teacher will also have time to do research, and every researcher will have time to teach in schools.

References

Althusser, L. (2008). *On ideology*. Verso Books.
Bachelard, G. (2002). *The formation of the scientific mind*. Manchester: Clinamen Press.
Baldino, R., & Cabral, T. (2013). The productivity of students' schoolwork: An exercise on Marxist rigour. *Journal for Critical Education Policy Studies, 11*(4), 1–15.
Boaler, J. (2010). *The elephant in the classroom. Helping children learn & love maths*. London: Souvenir Press.
Brenner, M. (1998). Meaning and money. *Educational Studies in Mathematics, 36*, 123–155.
Brown, T. (2011). *Mathematics education and subjectivity: Cultures and cultural renewal*. Dordrecht: Springer.
Clements, M. A. (2013). Past, present and future dimensions of mathematics education: Introduction to the third international handbook of mathematics education. In M. A. Clements, A. Bishop, C. Keitel, J. Kilpatrick, & F. Leung (Eds.), *Third international handbook of mathematics education* (pp. v–xi). New York: Springer.

Cobb, P. (2007). Putting philosophy to work: Coping with multiple theoretical perspectives. In F. Lester (Ed.), *Second handbook of research on mathematics teaching and learning* (pp. 3–38). Charlotte, NC: Information Age Publishing.

Evans, J. (1999). Building bridges: Reflections on the problem of transfer of learning in mathematics. *Educational Studies in Mathematics, 39,* 23–44.

Fendler, L. (1998) What is it impossible to *think? A genealogy of the educated subject.* In T. S. Popkewitz, & M. Brennan (Eds.), *Foucault's challenge. Discourse, knowledge, and power in education,* New York: Teachers College Press.

Foucault, M. (1979). Truth and power: an interview with Michel Foucault. *Critique of Anthropology, 4*(13–14), 131–137.

Foucault, M. (1997). The birth of biopolitics. In P. Rabinow (Ed.), *Michel Foucault, ethics: Subjectivity and truth* (pp. 73–80). New York: The New Press.

Gerofsky, S. (2010). The impossibility of 'real-life' word problems (according to Bakhtin, Lacan, Žižek and Baudrillard). *Discourse: Studies in the Cultural Politics of Education, 31*(1), 61–73.

Gutiérrez, R. (2010). The sociopolitical turn in mathematics education. *Journal for Research in Mathematics Education, 41,* 1–32.

Gutstein, E. (2003). Teaching and learning mathematics for social justice in an urban, Latino school. *Journal for Research in Mathematics Education, 23*(1), 37–73.

Jablonka, E., Wagner, D., & Walshaw, M. (2013). Theories for studying social, political and cultural dimensions of mathematics education. In M. A. Clements, A. Bishop, C. Keitel, J. Kilpatrick, & F. Leung (Eds.), *Third international handbook of mathematics education* (pp. 41–67). New York: Springer.

Jurdak, M. (2006). Contrasting perspectives and performance of high school students on problem solving in real world situated and school contexts. *Educational Studies in Mathematics, 63,* 283–301.

Klette, K. (2004). Classroom business as usual? (What) do policymakers and researchers learn from classroom research? In M. Høine & A. Fuglestad (Eds.), *Proceedings of the 28th Conference of the International Group for the Psychology of Mathematics Education* (Vol. 1, pp. 3–16). Bergen, Norway: University College.

Lacan, J. (1990). *Television.* New York: Norton & Company.

Lacan, J. (2007). *The other side of psychoanalysis: The seminar of Jacques Lacan book XVII.* New York: Norton & Company.

Lundin, S. (2012). Hating school, loving mathematics: on the ideological function of critique and reform in mathematics education. *Educational Studies in Mathematics, 80*(1), 73–85.

National Council of Teachers of Mathematics Board of Directors. (2016). *Mission.* Retrieved from http://www.nctm.org/About/.

Pais, A. (2012). A critical approach to equity in mathematics education. In O. Skovsmose & B. Greer (Eds.), *Opening the cage: Critique and politics of mathematics education.* Rotterdam: Sense Publishers.

Pais, A. (2013). An ideology critique of the use-value of mathematics. *Educational Studies in Mathematics, 84*(1), 15–34.

Pais, A. (2014). Economy: The absent centre of mathematics education. *ZDM—The International Journal on Mathematics Education, 46,* 1085–1093.

Pais, A. (2015). Symbolising the real of mathematics education. *Educational Studies in Mathematics, 89*(3), 375–391.

Pais, A. (2016). At the intersection between the subject and the political: A contribution for an ongoing discussion. *Educational Studies in Mathematics, 92*(3), 347–359.

Pais, A. (2017). The narcissism of mathematics education. In H. Straehler-Pohl, N. Bohlmann, & A. Pais (Eds.), *The disorder of mathematics education: Challenging the sociopolitical dimensions of research.* Switzerland: Springer.

Pais, A., & Valero, P. (2012). Researching research: Mathematics education in the political. *Educational Studies in Mathematics, 80*(1–2), 9–24.

Pais, A., & Valero, P. (2014). Whither social theory? *Educational Studies in Mathematics, 87*(2), 241–248.

Popkewitz, T. S. (2004). The alchemy of the mathematics curriculum: Inscriptions and the fabrication of the child. *American Educational Research Journal, 41*(1), 3–34.

Presmeg, N., & Radford, L. (2008). On semiotics and subjectivity: A response to Tony Brown's "Signifying 'students', 'teachers', and 'mathematics': A reading of a special issue". *Educational Studies in Mathematics, 69,* 265–276.

Radford, L. (2006). The anthropology of meaning. *Educational Studies in Mathematics, 61*(1–2), 39–65.

Radford, L. (2008). *Culture and cognition: Towards an anthropology of mathematical thinking.* In L. English (Ed.), Handbook of International Research in Mathematics Education (2nd ed., pp. 439–464). New York: Routledge, Taylor and Francis

Rancière, J. (1995). *Disagreement: Politics and philosophy.* Minneapolis: University of Minnesota Press.

Skovsmose, O. (2005). *Travelling through education: Uncertainty, mathematics, responsibility.* Rotterdam: Sense Publishers.

Skovsmose, O., & Valero, P. (2008). Democratic access to powerful mathematical ideas. In L. D. English (Ed.), *Handbook of international research in mathematics education* (2nd ed., pp. 415–438). New York: Routledge.

Skriabine, P. (2013). Science, the subject and psychoanalysis. *Psychoanalytical Handbooks, 27,* 51–54.

Sriraman, B., & English, L. (2010). Surveying theories and philosophies of mathematics education. In B. Sriraman & L. English (Eds.), *Theories of mathematics education: Seeking new frontiers.* Heidelberg: Springer.

Straehler-Pohl, H., & Pais, A. (2014). Learning to fail and learning from failure: Ideology at work in a mathematics classroom. *Pedagogy, Culture and Society, 22*(1), 79–96.

Tomšič, S. (2015). *The capitalist unconscious.* New York: Verso.

Utbildningsdepartementet. (2000). *Mathematics. Aims of the subject in upper secondary school.* Retrieved from http://www3.skolverket.se/. May 10, 2009.

Williams, J., & Wake, G. (2007). Black boxes in workplace mathematics. *Educational Studies in Mathematics, 64,* 317–343.

Žižek, S. (2012). *Less than nothing.* London: Verso.

Part III
Practices in the Sociopolitical in Mathematics Education

Teaching Financial Mathematics Through a Critical Approach in a University Environment

Celso Ribeiro Campos, Aurélio Hess and Renata Moura Sena

Abstract Financial education shares commonalities with the concept of education for citizenship particularly when it deals with social problems, for example family debts and/or irresponsible consumption. In this chapter, we present an approach, involving constructing a modelling activity in an undergraduate financial mathematics course, to connect financial education with mathematics education. In particular, critical mathematics education elements were incorporated in this activity to explore a critical financial education. Therefore, the aim of this chapter is to explore the affordances of a critical financial education through the adoption of a modelling pedagogy with roots in a democratic dialogic pedagogy, applied in the particular situation of Brazil. We observed an intense student involvement and great interest in the subject. Based on our experience, we claim to have achieved some financial education goals, as well as facilitated the students' participation in a critical discourse when debating themes they themselves proposed.

Keywords Financial education · Critical education · Mathematics education Financial mathematics · Mathematical modelling

1 Introduction

A concept that has long been commonplace among researchers on education and mathematics education, is an education for citizenship. For instance, Frankenstein (1989) places the teaching of mathematics within a logical foundation that links

C. R. Campos (✉) · A. Hess · R. M. Sena
FEA - Faculdade de Economia, Pontifícia Universidade Católica de São Paulo,
R. Monte Alegre, 984, Prédio Bandeira de Melo, 1o. andar, sala 122-A
Bairro Perdizes, São Paulo - SP 05014-901, Brazil
e-mail: crcampos@pucsp.br

A. Hess
e-mail: hessaurelio@gmail.com

R. M. Sena
e-mail: rmourasena@gmail.com

© Springer International Publishing AG 2018   113
M. Jurdak and R. Vithal (eds.), *Sociopolitical Dimensions of Mathematics Education*, ICME-13 Monographs, https://doi.org/10.1007/978-3-319-72610-6_7

education to a broader consideration of critical citizenship and social responsibility, focusing on how mathematics education could produce a critical citizenship, not just knowledge and awareness but how such knowledge enable critique of those in power and authority. In Brazil, both the Law of Directives and Bases of Education (Brazilian Ministério da Educação e Cultura 1996) and the National Curriculum Parameters (Brazilian Ministério da Educação e Cultura 2000), advocate, at all levels, an education for citizenship, in order to prepare students for an active, reflective and critical life, in which they can exercise their role as citizens aware of social, political, economic and environmental problems that permeate their society. In this context, we see financial education as a field to develop knowledge and information about personal finance that can help improve the life quality of people and their communities. The aim of this chapter is to explore the affordances of a *critical financial education* through the adoption of a modelling pedagogy with roots in a democratic dialogic pedagogy, applied in the particular situation of Brazil. Therefore, to clarify and deepen the discussion on the topic, we discuss the relationship between mathematics education and financial education and analyze interfaces between critical education and financial education. We present an educational project based on a mathematical modelling strategy, in order to put into practice the integration of these pedagogical aspects.

2 Financial Education and an Education for Citizenship

First, we expatiate upon financial education, focusing on some aspects that brings it close to an education for citizenship. Birochi and Pozzebon (2016, p. 268) have pointed out that "there is no single standard definition of the term financial education. Instead, there is a wide range of meanings and correlated terms". According to the authors, financial education can be broadly divided into two major streams, which they called instrumental and transformative (or critical).

As it can be seen in Fig. 1, the transformative stream is aligned with the idea of an education for citizenship. However, some approaches may combine aspects from both instrumental and transformative streams, such as the Organisation for Economic Co-operation and Development (OECD) publication, which states that financial education is:

> the process by which financial consumers/investors improve their understandings of financial products and concepts and, through information, instruction and/or objective advice, develop the skills and confidence to become more aware of financial risks and opportunities, to make informed choices, to know where to go for help, and to take other effective actions to improve their financial well-being. (OECD 2005, p. 26)

Others emphasize the transformative stream. Teixeira (2015, p. 13) points out that:

> Financial Education is not only to learn to economize, cut spending, saving and accumulating money, it is much more than that. It is seeking a better quality of life both today and in the future, providing the material security required for any unforeseen.

Financial education streams	Major underpinnings	Objectives	Authors
Instrumental	Financial education should promote efficiency and effectiveness of the financial system, through co-responsibility of the individuals (rights and liabilities). Individuals are consumers.	Financial education should act as a tool to improve the overall efficiency of the financial system, through training programs based on mastering of operational capabilities (knowledge about credit, debit, budget and negotiations).	Cole at al. (2009); Servon and Kaestner (2008); CGAP (2005).
Transformative or critical	Humanitarian and social approach. Individuals have huge socioeconomic constraints. Improvements are achieved by strengthening individual capabilities.	Financial education should aim at social-economic inclusion through strengthening of individual capabilities, targeting individual empowerment and social emancipation.	Cabraal (2011); Landvogt (2006); Sempere (2009); Johnston and Maguire (2005); Mayoux (2010); Augsburg and Fouillet (2010); Fernando (2006).

Fig. 1 Financial education streams. *Source* Birochi and Pozzebon (2016, p. 268)

Research carried out by OECD (2005) in developed and in emerging countries, pointed out a low level of financial awareness and, mainly, a lack of self-consciousness in vulnerable people. According to OECD, many people in different countries not only lacked the knowledge and skills needed to deal adequately with their personal finances but also ignored the very need for such knowledge. This led the OECD to recommend urgency in the implementation of government actions aiming to provide financial education to their population.

In response to the OECD's recommendation, the Central Bank of Brazil (BCB), which has been worried about the financial education of the Brazilian people, has developed a National Strategy for Financial Education (ENEF) to serve several purposes. Thus, BCB has created a Department of Financial Education, which has published a book containing some basic concepts, in order to inform people of some important aspects.

> Every citizen can develop skills to improve his/her quality of life and that of his/her family, improving behavioral attitudes based on personal finance management applied to his/her daily life. The Department of Financial Education of the Central Bank hopes that this copybook encourages you to make autonomous decisions regarding consumption, savings and investment, prevention and protection, considering your wishes and current and future needs. (BCB 2013, p. 3)

The BCB's publication deals with citizens' relationship to money, personal and family budget, the use of credit and debts administration, planned and conscious consumption, saving and investments, prevention and protection,[1] etc. According to BCB (op. cit.), financially well-educated consumers demand services and products that meet their needs, encouraging competition and playing an important role in monitoring the market, since they require greater transparency of financial institutions, contributing to the solidity and efficiency of the financial system. Besides these ideas, a good financial education can provide other important benefits, especially with regard to personal prosperity, self-esteem improvement and achievement of personal financial goals. Additionally, these benefits can spread from the personal to the family, and to communities.

Although BCB's attitude seems to be a good initiative, its extent and possible impact can be questioned. Thus, the National Strategy for Financial Education has elaborated a specific strategy to reach schools, both teachers and students. Along this line, a document was prepared for distribution to schools, containing orientations about financial education and presenting a conceptual model that aims to bring some principles that should guide the actions towards the desired future situation.

In this context, we list some important goals of financial education (adapted from Campos et al. 2015, p. 558) aligned with the transformative stream and which could be implemented in university financial math curriculum:

(i) Understand the basic functioning of the financial market and how interest rates affect the financial lives of citizens, for good or for bad, considering that people usually does not have deep awareness about how much interest rate really costs.

(ii) Practice conscious consumption, knowing and avoiding compulsive consumerism. Financial Education can help consumers not only concerning to their budgets, savings, and investments but also to control impulses associate to compulsive consumerism, to develop conscious consumption, associated with quality of goods and services, environment impacts, etc.

(iii) Be able to conveniently take advantage of the available funding opportunities.

(iv) Use credit consciously and wisely, seeking to avoid over-indebtedness. In other words, this means to learn how to use credits and how to maximize the funding access benefits.

(v) Understand the importance and benefits of planning and following up personal and family budget.

(vi) Understand the role of savings as a means to carry out projects and follow personal aims.

---

[1]Understand the financial risks and the preventive measures and appropriate protection for every situation.

Understand the importance of financial planning for retirement, how the national pension system is structured and what are the advantages and disadvantages of adopting independent strategies, being the manager of your own investments. (BCB 2013, p. 9)

(vii) Help to disseminate good financial practices among family and friends.
(viii) Develop a culture of prevention that is, planning for the future considering the mishaps that can happen.
 (ix) Be able to organize and keep a good personal financial management.
  (x) To make a retirement plan, considering that life expectancy has increased and people spend more time in retired condition.

In summary, financial education promises to help citizens to be more conscious of their consumption habits. The organization and discipline required for the practice of financial education may lead to better and more efficient decisions on using one's scarce financial resources. Financial education could transform consumption habits and reduce impulse purchases, bringing control and rationality to consumers.

3 The Impact of Economic Changes in Brazil on Citizens' Financial Behavior

To comprehend the importance of financial education in Brazil, it is necessary to understand the recent history of Brazil's economic scenario. The high inflation that occurred in Brazil during the 1980s and middle of 1990s created important scars in consumer behavior. At that time, it was common for people to focus on spending their money just on the day they received it because the money would lose value daily. According to the BCB, the National Consumer Price Index reached 82% in just one month in March 1990. This index captures the inflation, which precisely affects the low-income population the most.

Economic instability leads people not to formulate long-term expectations, because the inflationary cycle creates uncertainty, even the short run. Thus, there was a generalized preference for liquidity among agents, because it was very necessary to guarantee basic daily consumption.

In the period between 1985 and early 1994, Brazil has experienced six economic plans (*Cruzado*, *Cruzado 2*, *Bresser*, *Verão*, *Collor* and *Collor 2*) and three currency exchanges (*Cruzado*, *Cruzado Novo* and *Cruzeiro Real*), whose economic stabilization attempt has failed. Nevertheless, in mid-1994, a new economic stabilization plan came into force, with a new currency (*Real*), which was finally successful.

The so-called *Plano Real* has pushed prices to stability and has opened the domestic market, leading to increasing imports of goods for supply. This plan substantially reduced the *inflationary tax*, which prejudiced poor people. The price predictability, the employment and income growth brought with it demands from a huge number of people who were outside the market (Fortuna 2008).

From the low-income consumers' point of view, the *Plano Real* had meant access to unthinkable goods and services, like yogurts, meat, and dentists. The plan became known as *yogurt plan* because of the earlier high prices, which limited access to it, or *chicken plan*, because with just R$1,00 people could buy 1 kg chicken meat. Consumption increased by 40% (G1 2014).

After the economic stabilization, some typical behaviors from consumers and savers became clearly determined because of the predictability of income, interest, access to the banking system and other factors. From the perspective of microeconomic theory, consumption is a function of income and prices. With prices controlled and increasing income, Brazilian consumers spent a lot of money on several types of products in order to satisfy a huge pent-up demand.

According to de Ferreira (2008), the consumer's decisions reflect the immediate satisfaction possibility, as it seeks to guarantee a present pleasure. From this point of view, consumers choose the present pleasure, instead of saving for buying in the future, because of their inflationary period memory and, fundamentally, because of their not knowing financial control.

One of the reasons that Brazilian consumers do not have the habit to control their budget and the consumption is their memories of the inflation period. However, another reason is the increase in income in recent past years. According to Brazilian Geographic and Statistics Institute, between 2002 and 2015, real average income has increased 14%, minimum wage grew 44%, while families' consumption grew 75%.

Figure 2 shows the growth rates of four variables: minimum wage, average income, GDP *per capita* and household consumption; and shows that consumption grew above income in the period. According to the Keynesian concept of marginal propensity to consume, an extra income causes a variation on spend in consumption and Brazilian consumers confirmed this theory responding greatly and quickly to income variations.

In addition, during the same period, Brazil improved the income's transfer[2] for poor population, e.g., *Bolsa Família*, which is a government program benefiting low-income families. This program had beneficiated 13.8 million households, embracing 26% of the population in 2012 (United Nations 2015). The budget of this program corresponded to 0.53% of GDP in 2013 and was fully financed by the social security budget. According to the United Nations' Human Development Report (2015), since the program was launched, Brazil reduced poverty by about 8 percentage points. This introduced more consumers to the economy, people who could not be included before because of poverty.

Another important point related to income and consumption was the decrease in unemployment rates. According to IBGE (2017), Brazilian GDP grew, in real terms by 49.5% and one of the consequences was a reduction on inoccupation rates, e.g., in 2002 the rate was 10.5% and reached 4.3% in 2014, increasing to 6.9% in 2015 (Fig. 3). The employment supply increase has increased income for a population who then transform this reality into consumption.

---

[2]Income transfer or cash transfer comprehends government initiatives in order to destine money resources from rich people to vulnerable people, which means to use money from taxes and tributes to pay a monthly amount to poor families. See Medeiros et al. (2007).

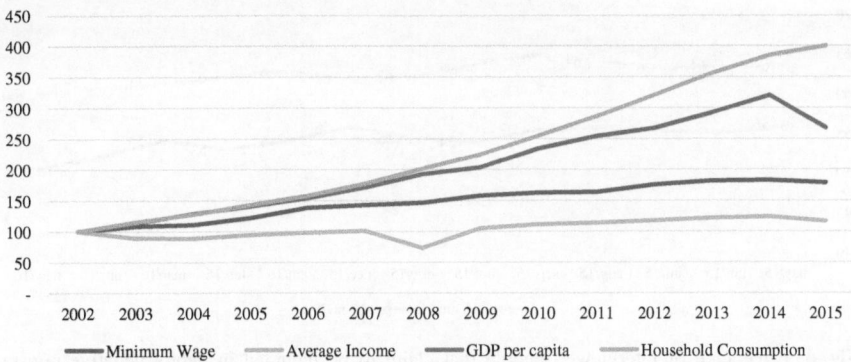

Fig. 2 Growth rates of minimum wage, average income, GDP per capita and household consumption, between 2002 and 2015 (2002 = 100). *Source* Data from Instituto Brasileiro de Geografia e Estatística (IBGE 2016)

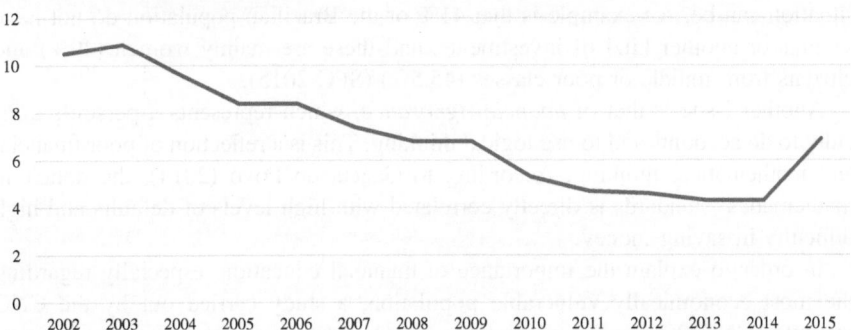

Fig. 3 Inoccupation rate, between 2002 and 2015 (%). *Source* Data from Instituto Brasileiro de Geografia e Estatística (PME IBGE 2016)

The abundance of money led people who were out to get in consumption (entry) and those who were into enlarge their consumption standard. This led to some problems. According to research carried out by Credit Protection Service (SPC), more than 70% of Brazilian people have an incorrect perception about debts. For more than 50%, to have debts meant *just delaying paying of accounts*, which is a wrong perception (Valor Econômico 2015).

Moreover, another study from SPC (2015) concluded that 48% of those interviewed do not have any personal budget control. Around 60% reported much difficulty in controlling monthly income and expenditure, and 33% appealed to credit, like credit cards and banking account limits. According to this study, Brazilian consumers have a relatively poor knowledge and attitude related to financial education.

People do know what they would have to do to reach a financial balance, but do not do anything towards reaching such balance. One way of explaining this

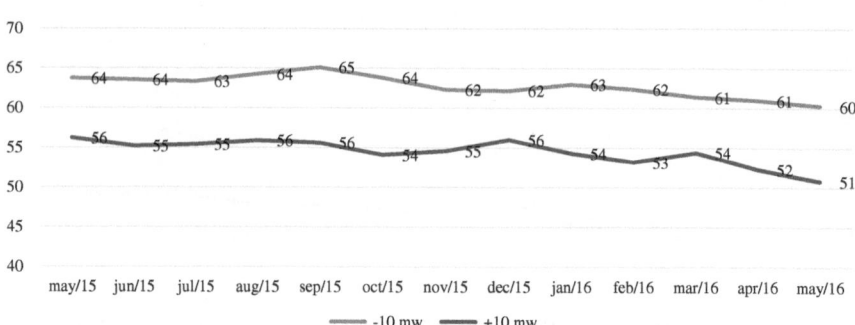

Fig. 4 Percentage of interviewed people that claim to be included in debt's negative register minimum wage reference (%). *Source* CNC (2016)

behavior is to relate it to financial subject unfamiliarity, lack of habits, lack of discipline and difficulty in seeing how beneficial a good and healthy financial situation can be. An example is that 41% of the Brazilian population do not have savings or another kind of investments, and these are mainly women (49%) and citizens from middle or poor classes (45.5%) (SPC 2015).

Another issue is that of *financial ignorance*, which represents a person's difficulty to do accounts and to use logical thinking. This is a reflection of poor financial and mathematical thinking. According to Gazeta do Povo (2014), the deficit in mathematics' standards is directly correlated with high levels of defaults and high difficulty in saving money.

In order to explain the importance of financial education, especially regarding the most economically vulnerable population, a study carried out by the CNC institute (CNC 2016) showed that the people who live under the 10 minimum wage[3] are the most affected by the default problem, as can be seen in Fig. 4.

This research (CNC 2016) has also revealed that the debts are mainly from credit cards, car leasing, hypothec, insurance, etc. Another search carried out by *Serasa Experian* shows that Brazil's north region, which is the poorest region, has the highest percentage of families declared with debts, at 31.1%. Nationally speaking, among people between 31 and 35 years, 29.3% have been declared to be in debt (Serasa Experian 2016). In addition, people's profile revealed that 23% of those defaulting are among young adults living in poor urban areas, as can be seen in Fig. 5.

The economic and political crises that Brazil confronts nowadays, with more than 6% reduction in GDP between 2015 and 2016, and the increase of unoccupied rates (Fig. 3), brings people to renegotiate their debts, control compulsive consumerism and look for ways to save money for emergencies. Despite the financial crisis, these numbers reveal an alarming situation, especially for the most

---

[3]A minimum wage in 2016 were of R$880.00 or around US$260.00.

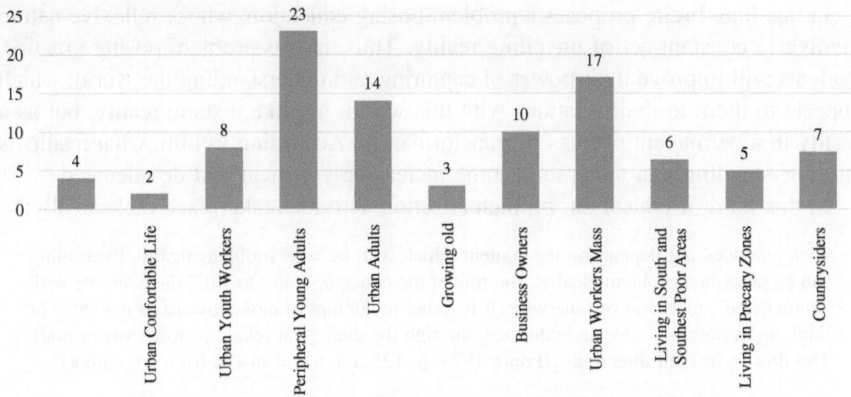

Fig. 5 Percentage of people in debt profile, 2014. *Source* Serasa Experian (2016)

vulnerable population, which perhaps could be less alarming if people had more orientation on how to deal with their personal finances or family budgets.

In summary, spreading financial education concepts could potentially improve people's quality of life as much as the economy as a whole. Practicing conscious consumption, planning personal budgets, saving money for buying something in the future or for retirement tend to help economy to become healthier and richer.

4 A Critical Financial Education

The pedagogical approach that we have developed towards an understanding of some important concepts related to financial education, aligned with the transformative stream, is conceived under the critical education proposals.

The Brazilian educator Paulo Freire substantially contributed to a better foundation of a critical theory of school learning. He emphasized the basis of a true democratic pedagogy, which fights against authoritarian relations through dialogue. The special conditions of Latin America during the 60s and 70s marked his work, but his work remains current until today.

According to Freire (1973), education must have a constant attempt to change attitude, replacing old passivity habits by new habits of participation and interference within student's reality. Campos (2016) points out that Freire shows that the attitude of a critical and criticizing education should lead people to a new position facing the problems of their time and space.

As for Freire, a critical educational science is a process of awareness as it is "[…] the process in which people, not as recipients, but as knowing subjects, achieve a deepening awareness both of the sociohistorical reality which shapes their lives and of their capacity to transform that reality" (Freire 1970, p. 27).

In this line, Freire proposes a problem-posing education, whose reflexive nature involves a constant act of unveiling reality. Thus, in this problem-posing practice, students will improve their power of capturing and understanding the world, which appears to them in their relations with this world, not like a static reality, but as a reality in a permanent process of transformation. According to him, what results is an understanding that tends to become increasingly critical and de-alienated.

In this kind of education, problematization activities take place dialectically:

> Dialogue does not depend on the content which is to be seen problematically. Everything can be presented problematically. The role of the educator is not to "fill" the educatee with "knowledge", technical or otherwise. It is rather to attempt to move towards a new way of thinking in both educator and educatee, through the dialogical relationships between both. The flow is in both directions. (Freire 1973, p. 125, quotation marks from the author)

In short, Freire presents a democratic pedagogy based on dialogue, through an active, critical and criticizing way. Thus, an education based on problematization of the contents, which are presented as relevant to the learners, challenging. This praxis results in a process of reflection-action by the student on his world/reality, activating his/her awareness from the generating themes.

> For only as man grasp the themes, can they intervene in reality instead of remaining mere onlookers. And only by developing a permanently critical attitude can men overcome a posture of adjustment in order to become integrated with the spirit of the time. (Freire 1973, pp. 5–6)

5 Features of a Proposed Critical Financial Education Pedagogical Strategy

Skovsmose (2014), a seminal writer in critical mathematics education, emphasizes Freire's idea of dialogue in the relationship between teacher and students. For him, it is important to break down the figure of the knowledge-owner-teacher and takes effect the presence of the one who teaches and is taught in a dialectical relationship with the students, who become co-responsible for an educational process in which all grow.

> The ideas concerning the dialogue and the student-teacher relationship are developed from the general point of view that education must belong to a process of democratization. If a democratic attitude is to be developed through education, education as a social relationship should not contain fundamentally undemocratic features. It is not acceptable that the teacher (alone) has the decisive and prescribing role. Instead the educational process must be understood as a dialogue. (Skovsmose, op. cit., p. 350)

According to Skovsmose (2005, p. 114), "Mathematics education might serve a further development of a concern for democracy and ensure social inclusion. It might, however, provoke exclusion as well. This leads me to consider the importance of critical mathematics education".

In addition, Skovsmose highlights that an important aspect of critical mathematics education is problem orientation in the teaching-learning process. To select the types of problems for teaching, one should take into account what is really relevant to the student and it must have a close relation to objective existing social problems.

According to Alrø and Skovsmose (2004), critical mathematics education is an approach which values certain mathematics learning qualities.

> Critical mathematics education is concerned with the way mathematics in general influences our cultural, technological and political environment and with the purposes for which the mathematical competence should serve. [...] The critical mathematics education is also concerned with issues such as *how learning mathematics can support the development of citizenship* and *how the individual can be empowered through Maths*. (Alrø and Skovsmose 2004, p. 16, emphasis in original)

There are many ways or strategies to carry out critical mathematics education. The *thematization* is more focused on primary and secondary education, while the *organization-in-projects* is more conducive to college education. Skovsmose (2014) does not consider them sufficient or ideal, but just reasonable, and emphasizes, as the most effective strategy, the *problematization*. For it to work as a practical mechanism for critical mathematics education, it is important that students understand the relevance of the problem, which should be related to their experience. Problems should be linked to processes important for society in general and when assuming responsibility for solving them, students must design a political and social engagement.

That said, as practitioners of a critical approach to education, we propose a pedagogical experiment concerning financial education. In mathematics, there are many content topics related to students' daily lives, including financial mathematics, which we believe is a key link to involve the practice of education for citizenship, in financial education and mathematics contents.

In our pedagogical strategy for a critical financial education, we use mathematical modelling. Generally, one may create a model for interpreting and studying natural or social phenomena. The advancement of technology has made the use of virtual models quite common, and these allow a great quantity of simulations. The objective of creating a model can be analytical, explanatory, pedagogical, for forecast, etc. Mathematics is particularly abundant in allowing model creation, when dealing with quantitative variables.

> In this perspective, a set of symbols and mathematical relations, which aims to translate a phenomena or a problem from a real situation, is called mathematical model. (Biembengut and Hein 2003, p. 12)

The process that involves obtaining a mathematical model is known as mathematical modelling. Modelling is similar to an art, when creating models for different purposes, and can be seen as a form of creation and expression of knowledge.

For us, mathematical modelling is a method (or pedagogical strategy) that can be employed at various school levels, from elementary mathematics to graduate level. It can be conceptualized as a learning environment to be built in the classroom in which students are asked by the teacher to investigate, through mathematics, situations extracted from daily life or even other sciences.

> The mathematical modelling process can be a way to awaken in students the interest in mathematical content, to the extent that they have the opportunity to study, through various investigations, situations that have practical application and value their critical sense. (Campos et al. 2011, p. 47)

Complementing this idea, Campos et al. (2015) have pointed out that teaching concepts of financial mathematics by themselves is not enough to accomplish the objective of forming citizens and promoting critical financial education if they were not contextualized in real or realistic situations, close to the student's life.

Nevertheless, a link between financial education and critical mathematics education can be created, as long as some discussions and debates are to be carried out in order to develop a deeper awareness along with social and economic issues that affect students and their families, friends and communities. A critical consciousness should also include discussions over government responsibilities regarding the country's financial system, its regulation and its decisions, which affect all citizens. This is precisely what we want to call critical financial education.

Critical education aligns itself with the idea of education for critical citizenship, to the extent that it incorporates the tensions and contradictions between what is and what should be in a democratic society that is grounded on equality, freedom and justice. Thus, critical mathematics education reminds us of a social character of pedagogical work, which besides seeking to give meaning to the mathematical content, seeks to do so in a democratic way, encouraging students to develop critical thinking, ethical responsibility and political awareness.

Moreover, Valero (2015) has observed students' decreasing engagement in mathematics, fundamentally caused by a gap between the forms of subjectivity promoted by mathematics as an area of schooling and the forms of subjectivity experienced by them in their everyday life. Thus, we interpret that what Valero stated reinforces the need for approaching mathematical content in a way that assists students to realize its importance and are encouraged to engage themselves in their pedagogical environment, which is precisely what critical mathematics education advocates. Therefore, a critical financial education is in line with the purposes of critical mathematics education, insofar as it aims to bring to the classroom, a discussion about the social problems arising from the mismanagement of personal finances, and aims for a transformation of the harsh reality exposed by the alarming data presented in relation to household indebtedness and excessive consumerism.

6 Example of a Critical Pedagogical Activity in Financial Education

6.1 Description of the Activity

For the purposes of putting into practice the idea of a critical financial education through a modelling strategy, we designed a pedagogical activity, which addressed selected financial education concepts in a financial mathematics class, carried out by the first author in an undergraduate course. We proposed a student research project, focusing on basic ideas from financial education.

The 20 students organized themselves into four groups and chose the following topics:

(i)   Brazil's basic interest rates;
(ii) Banking interest rates for loans and banking profits;
(iii) Brazil's budget: debts and incomes;
(iv) Family budget.

Students were required to make a report and do a presentation on the selected topic. They agreed that a period of two weeks would be sufficient for the task. All presentations would be done in sequence at the same day, and each one of them should be followed by a discussion where everyone should participate.

The first group explained the BCB's basic interest rate (SELIC) and showed its historical series (Fig. 6).

The group reported that considering a 9.2% expected annual inflation (IBRE 2015), the real interest rate would be of 4.62%, which was the world's highest interest rate at that time (Moneyou 2015), since the SELIC was of 14.25% per year. It was also explained that the SELIC rate affects banking interest rates for consumer's loans and, as it presents an upward trend, the perspective for the loan borrowers was not good. The discussion that followed the presentation was mainly about whether government (or BCB) should increase the interest rates or not.

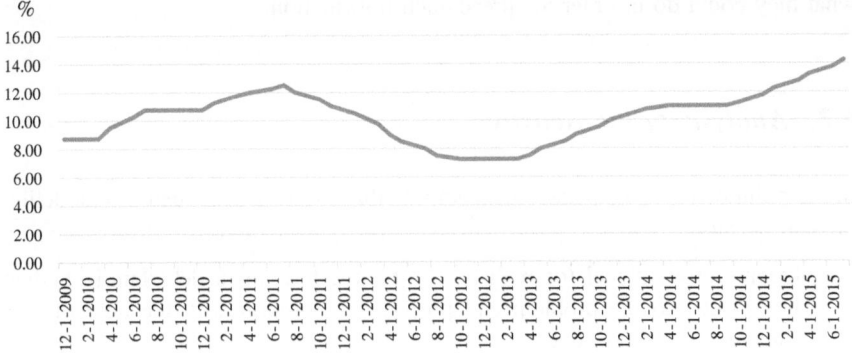

Fig. 6 SELIC historical series. *Source* Data from BCB (2015, in students' report)

The second group showed data (BCB 2015) revealing that the average interest rate for personal loans (hot money) from the six greatest Brazil's banks was of 11.0% per month, which represents approximately 250% annual rate. As for credit cards, the same banks had even higher interest rates, that is, 12.6% per month, or 316.7% annual rate (average). The group also showed data (Costas 2015) revealing that the Brazil's greatest private bank (Itaú) had obtained profits of R$20.6 billion in 2014, which represented an increase of 30.2% when compared to the year before. In addition, quoting *Economática Consulting* (apud Costas, op. cit.), they revealed that the profitability of Brazil's banks was of 18.23% in 2014, more than twice the profitability of US banks (7.68%). The debate that followed this presentation had two main subjects: how can Brazilian banks improve their profits along with an economic recession scenario and how people should avoid banking loans.

The third group showed data from BCB, revealing that in 2013, the national account has had a nominal deficit of R$173.8 billion (BCB 2014), i.e., the outcomes overcome the incomes by about this amount, which represented 3.25% of GDP. They also showed preliminary data from 2014, revealing a primary deficit of R$32.5 billion and a nominal deficit of R$343.9 billion, which represented 6.70% of GDP (Exame 2015). After the presentation, students discussed the bad example given by central government, which could not eliminate the country's deficit, and which causes vulnerability in controlling and combating inflation, resulting in an upward trend.

The fourth group presented a fictional family budget, detailing all debts. Based on their own families' expenses, they have built a long list of consumed items and a supposed income, which was not enough to overcome the expenses. Thus, the family had a deficit and the group wanted to discuss how to eliminate it. The question raised in the debate was on how can a family reverse a deficit if they do not make a detailed control like this. They also discussed the need for some items, which some students considered unnecessary or superfluous, especially for an indebted family.

After that, the teacher addressed some important financial education issues like budget planning, excessive consumption and misuse of credit cards, among others. Students pointed out the need for awareness of these concepts and have argued what they could do in order to spread such information.

6.2 Analysis of the Activity

In our point of view, the pedagogical activity that we have described is a modelling project, because:

(a) Students brought real (or realistic) problems, like an in debt family budget. They discussed the problem and proposed a solution with the help of their colleagues and the teacher.

(b) Students also brought other problems, which were concerned with the interest rate from private Brazilian banks. They argued about some causes for this issue and showed that there was a poor handing of the country's economic policy, besides the greed of banks for profits, no matter who is harmed.

(c) In addition, the students were the ones who chose the themes.

The activity is inserted in the context of critical mathematics education, as the students faced a real social economic problem, which affects everyone, but especially the disadvantaged and people that are more vulnerable. Many discussions were carried out concerning this financial problem and the students concluded that a dangerous lack of information was greatly responsible for the problems. Otherwise, they have strongly criticized the government's conduct and that of one of the banks. In facing the problems, they also reflected on what actions they could take to help solve the problem of a lack of information.

In this proceeding with the activity, we have created a democratic environment, to the extent that the students chose the themes and had a voice during the presentations and debates carried out in the classroom. The students engaged in dialogue, discussing and reflecting on solutions to a serious social problem. In other words, this activity seems to support the development of citizenship and students felt empowered through the engagement on this modelling project.

As was pointed out in the fourth group presentation, budget planning is important for any family, in order to administer their finances. The prejudice caused by unnecessary or superfluous consumption was also pointed out. In this line, the debates argued for a qualitatively better world, as long as the students have discussed the importance of spreading information about some financial education concepts. Altogether, bringing public economic policies to the debate, we have performed what. Skovsmose called *problematization*. Insofar, students understood the relevance of the problem, which is important for society, and designed a political and social engagement in the debates, thus performing the critical mathematics education.

Moreover, we were able to fight some of the problems that Valero (2015) has stated, namely addressing the gap between the forms of subjectivity promoted by mathematics as an area of schooling and the forms of subjectivity experienced by them in their everyday life, besides the need of approaching mathematical contents linked to reality.

As we have pointed out, for Skovsmose (2014), it is important that students understand the relevance of the problem, which should be related to their experience. In addition, he said that the problems should be linked to processes important for society in general and when assuming responsibility for solving them, students must design a political and social engagement. That is precisely what happened in this pedagogical activity, the students brought the financial problems, whose relevance can be seen through the data that we have shown, that is, the great number of people dealing with debt troubles.

Concerning Freire's ideas (1970, 1973), we have engaged a democratic pedagogy based on dialogue, through an active, critical and criticizing way. This kind of pedagogy, based on problematizing the contents, which were presented as relevant and challenging to the learners, have resulted in a process of reflection-action by the students on their world and reality, boosting their awareness from generating themes, exactly as Freire has pointed out.

Finally, financial education was also carried out in this pedagogical activity, as students have discussed some of its issues, e.g., family budgets, debts, consumerism, etc. Consonant to the fact that we have done a link between financial education and critical mathematics education, we can say that this pedagogical activity played what we called critical financial education.

6.3 Student Learning Outcomes

In this pedagogical activity, we realized that students showed a great involvement in all steps. The debates and discussions over the problems raised from the presentations were noticeable and the engagement of the students was also noteworthy.

We were able to see a critical consciousness emerging on the discussions over government lack of responsibility in managing the country's budget.

Nevertheless, it is relevant to note that some stressing issues played a minor role in this pedagogical approach. Students showed some difficulties in drawing reports, revealing a weak performance caused by their unfamiliarity with this kind of task. In addition, some discussions had to be abbreviated due to the short time available for the accomplishment of the project.

7 Concluding Remarks

The aim of this project was to carry out an activity involving all students in a discussion over some financial education issues within a financial math class, developing a critical approach under a mathematical modelling strategy. There was an intense involvement of the students in all presentations, revealing their interest in the subject. Thus, goals (i) and (v) listed for financial education were assigned in the project, namely: understand the basic functioning of the financial market and how interest rates affect the financial lives of citizens, for good or for bad; and understand the importance and benefits of planning and following up personal and family budget.

The critical approach was mainly revealed during the discussions, when students showed a strong indignation about the issues pointed out by the groups. Many criticisms were about government attitudes, when failing on the management of its accounts and increasing the basic interest rates. The banks were criticized too, as they punish people with high interest rates, while increasing their profits in an economic recession scenario.

Moreover, the last group brought a realistic example closer to the core of the transformative stream of financial education. Their presentation, along with the following debates, leaded students to realize and feel the importance of taking care of their personal and families' finances.

From our perspective, financial education is of great importance for all citizens and schools should encourage its insertion in mathematics classes at all levels. In addition, a critical education approach mediated by a mathematical modelling strategy has revealed to be worthy for achieving financial education goals in both a schooling and university environment.

Based on what we have experienced with this pedagogical activity, it seems important to connect financial education within financial mathematics classes and curricula, so that students could have a chance to experience its concepts, which are aligned with the idea of education for citizenship.

References

Alrø, H., & Skovsmose, O. (2004). *Dialogue and learning in mathematics education: Intention, reflection, critique.* Dordrecht: Kluwer Academic Publishers.

Augsburg, B., & Fouillet, C. (2010). Profit empowerment: The microfinance institution's mission drift. *Perspectives on Global Development and Technology, 9*(3–4), 323–351. https://doi.org/10.1163/156914910x499732.

BCB. (2013). *Caderno de Educação Financeira—Gestão de Finanças Pessoais (Conteúdo Básico).* Brasília: Banco Central do Brasil.

BCB. (2014). *Special data dissemination standard.* Brasília: Banco Central do Brasil. http://www.bcb.gov.br/pec/sdds/ingl/sddsi.htm. Accessed March 20, 2015.

BCB. (2015). *Interest rates.* Brasília: Banco Central do Brasil. http://www.bcb.gov.br/?INTEREST. Accessed April 30, 2015.

Biembengut, M. S., & Hein, N. (2003). *Modelagem matemática no ensino* (3rd ed.). São Paulo: Contexto.

Birochi, R., & Pozzebon, M. (2016). Improving financial inclusion: Towards a critical financial education framework. In *Revista de Administração de Empresas* (pp. 266–287). São Paulo: FGV/EAESP.

Brazilian Ministério da Educação e Cultura. (1996). *Lei de Diretrizes e Bases da Educação Nacional.* Brasília: MEC/SEF.

Brazilian Ministério da Educação e Cultura. (2000). *Parâmetros curriculares nacionais.* Brasília: MEC/SEF.

Cabraal, A. (2011). *The impact of microfinance on the capabilities of participants* (Ph.D. thesis). School of Economics, Finance and Marketing, RMIT University. Melbourne, Austrália.

Campos, C. R. (2016). *Towards critical statistics education.* Saarbrücken, Germany: Lambert Academic Publishing.

Campos, C. R., Teixeira, J., & Coutinho, C. Q. S. (2015). Reflexões sobre a educação financeira e suas interfaces com a educação matemática e a educação crítica. *Educação Matemática Pesquisa, 17*(3), 556–577. São Paulo: PUC-SP.

Campos, C. R., Wodewotzki, M. L. L., & Jacobini, O. R. (2011). *Educação estatística—Teoria e prática em ambientes de modelagem matemática.* Belo Horizonte: Autêntica.

CGAP—Consultative Group to Assist the Poor. (2005). *Annual report.* Washington, DC, USA: CGAP.

Cole, S., Sampson, T., & Zia, B. (2009). *Valuing financial literacy training*. Washington, DC, USA: World Bank.

CNC, Confederação Nacional do Comércio de Bens e Serviços. (2016). *Pesquisa Nacional de Endividamento e Inadimplência do Consumidor (PEIC)—Maio, 2016*. http://cnc.org.br/sites/default/files/arquivos/graficos_peic_maio_2016.pdf. Accessed February 21, 2017.

Costas, R. (2015). *Por que os bancos brasileiros lucram tanto?* BBC Brasil. http://www.bbc.com/portuguese/noticias/2015/03/150323_bancos_lucros_ru. Accessed February 15, 2017.

de Ferreira, V. R. M. (2008). *Psicologia Econômica: como o comportamento econômico influencia nas nossas decisões*. Rio de Janeiro: Elsevier.

Exame. (2015). *Brasil fecha 2014 com déficit primário*. São Paulo: Editora Abril. January 30, 2015. http://exame.abril.com.br/economia/noticias/brasil-fecha-2014-com-deficit-primario-de-r-32-536-bilhoes. Accessed March 11, 2015.

Fernando, J. (2006). Microcredit and empowerment of women: Visibility without power. In J. Fernando (Ed.), *Microfinance perils and prospects* (pp. 187–237). New York, USA: Routledge.

Fortuna, E. (2008). *Mercado Financeiro: Produtos e serviços* (17th ed.). Rio de Janeiro: Qualitymark.

Frankenstein, M. (1989). *Relearning mathematics: A different third R-radical math(s)* (Vol. 1). London: Free Association Books.

Freire, P. (1970). *Cultural action for freedom*. Cambridge, MA: Center for the Study of Social Change.

Freire, P. (1973). *Education for critical consciousness*. New York: Continuum.

G1. (2014). *Frango e iogurte são símbolos do Plano Real*; *veja dez curiosidades*. July 01, 2014. http://g1.globo.com/economia/noticia/2014/07/frango-e-iogurte-sao-simbolos-do-plano-real-veja-dez-curiosidades.html. Accessed February 17, 2017.

Gazeta do Povo. (2014). *"Ignorância Financeira" afeta economia*. July 03, 2014. http://www.gazetadopovo.com.br/economia/ignorancia-financeira-afeta-economia-eafv8uhsria2xwgm1quj9ca4u. Accessed February 20, 2017.

IBGE. (2016). Instituto Brasileiro de Geografia e Estatística. Rio de Janeiro. http://www.ibge.gov.br/home/estatistica/indicadores/trabalhoerendimento/pme_nova/defaulttab_hist.shtm. Accessed November 20, 2016.

IBGE. (2017). Instituto Brasileiro de Geografia e Estatística. Rio de Janeiro. https://ww2.ibge.gov.br/home/estatistica/indicadores/pib/defaulttabelas.shtm. Accessed October 05, 2017.

IBRE. (2015). *Expectativa de inflação dos consumidores*. São Paulo: FGV. http://portalibre.fgv.br/lumis/portal/file/fileDownload.jsp?fileId=8A7C82C54DB5CA9F014E2909829E0B92. Accessed January 25, 2016.

Johnston, B., & Maguire, T. (2005). *Adult numeracy: Policy and practice in global contexts of lifelong learning*. Working Papers. Melbourne, Australia: ALNARC, Victoria University. Retrieved from http://www.voced.edu.au/content/ngv%3A27913. Accessed June 10, 2017.

Landvogt, K. (2006). *Critical financial capability. Financial literacy, banking and identity conference*. Melbourne, Australia: RMIT University.

Mayoux, L. (2010). Reaching and empowering women: Towards a gender justice protocol for a diversified, inclusive, and sustainable financial sector. *Perspectives on Global Development and Technology, 9*(3–4), 581–600. https://doi.org/10.1163/156914910x499822.

Medeiros, M., Britto, T., & Soares, F. (2007). *Transferência de renda no Brasil*. São Paulo: CEBRAP. http://www.scielo.br/scielo.php?script=sci_arttext&pid=S0101-330020070003000 01. Accessed August 15, 2017.

Moneyou. (2015). *Ranking de juros reais*. São Paulo: Infinity Asset Management. http://moneyou.com.br/wp-content/uploads/2015/07/rankingdejurosreais280715.pdf. Accessed December 8, 2017.

OECD. (2005). *Improving financial literacy: Analysis of issues and policies*. Paris: Secretary-General of the OECD.

PME IBGE. (2016). *Pesquisa Mensal de Emprego*. Rio de Janeiro: Instituto Brasileiro de Geografia e Estatística. ftp://ftp.ibge.gov.br/Trabalho_e_Rendimento/Pesquisa_Mensal_de_Emprego/fasciculo_indicadores_ibge/2016/. Accessed November 15, 2016.

Sempere, M. C. (2009). Evaluación de la alfabetización: El proceso de crear un marco de evaluación para el enfoque REFLECT. International reflections on issues arising from the benchmarks and call for action. *Revista EAD—Educación de Adultos y Desarrollo*. Germany: CVV Intertational. Retrieved from https://www.dvv-international.de/es/educacion-de-adultos-y-desarrollo/ediciones/ead-712008/reflexiones-internacionales-sobre-aspectos-derivados-de-los-puntos-de-referencia-y-del-llamamiento-a-la-accion/evaluacion-de-laalfabetizacion-el-proceso-de-crear-un-marco-de-evaluacion-para-elenfoque-reflect/. Accessed June 10, 2017.

Serasa Experian. (2016). *Estudos de Inadimplência*. https://www.serasaexperian.com.br/estudo-inadimplencia/. Accessed February 21, 2017.

Servon, L., & Kaestner, R. (2008). Consumer financial literacy and the impact of online banking on the financial behavior of lower-income bank customers [Special issue: Financial literacy: Public policy and consumers self-protection]. *Journal of Consumer Affairs, 42*(2), 271–305. https://doi.org/10.1111/j.1745-6606.2008.00108.x.

Skovsmose, O. (2005). Ghettorising and globalisation: A challenge for mathematics education. *Zetetiké, 13*(24), 113–142. Campinas, SP: UNICAMP—Faculdade de Educação.

Skovsmose, O. (2014). *Critique as uncertainty*. Charlotte, NC, USA: Information Age Publishing.

SPC Brasil. (2015). *Indicadores econômicos SPC Brasil e CNDL*. https://www.spcbrasil.org.br/pesquisas/pesquisa/2453. Accessed November 15, 2016.

Teixeira, J. (2015). *Um estudo diagnóstico sobre a percepção da relação entre educação financeira e matemática financeira*. Thesis, Doctorate degree in Mathematics Education. São Paulo: PUC-SP.

United Nations. (2015). *Human development report—Work for human development*. UNDP. http://hdr.undp.org/sites/default/files/2015_human_development_report.pdf. Accessed February 21, 2017.

Valero, P. (2015). Re-interpreting students' interest in mathematics: Youth culture and subjectivity. In U. Gellert, J. Giménez, C. Hahn, & S. Kafoussi (Eds.), *Educational paths to mathematics* (pp. 15–32). New York: Springer.

Valor Econômico. (2015). *Mais de 70% dos brasileiros têm noção errada do que é estar endividado*. São Paulo, March 18, 2015. http://www.valor.com.br/financas/3961686/mais-de-70-dos-brasileiros-nao-sabem-o-que-e-estar-endividado. Accessed February 20, 2017.

Mathematics Education for Social Justice: A Case Study

Gustavo Bruno, Natalia Ruiz-López and César Sáenz de Castro

Abstract In this chapter, we present a case study of a mathematics teacher in a school in Madrid who identifies himself as an educator for social justice. We analyze both his classroom practice using an observation protocol, and key elements in his biography through a biographic-narrative interview, to compare what the teacher declares as ideal with his action in the classroom. From the observations of his practice, it seems that the teacher adopts an instrumental and value-neutral perspective on mathematics, but from the biographic-narrative interview we obtain both an understanding of the origins of this apparent instrumental perspective, and also many notable intuitions and critical perspectives on sociopolitical issues related to mathematics, teacher training and mathematics education research.

Keywords Mathematics education · Social justice · Teacher training
Secondary level · Classroom practice

1 Introduction

The economic crisis in Spain is causing an increasing sharp gap between the people who have more and those who have less, and an increase of poverty and exclusion pockets, unknown for decades. This situation is, of course, affecting the schools, and in a very short time the Spanish education system has developed serious issues of educational inequity. The multidisciplinary group to which the authors belong, *Educational Change for Social Justice (GICE)* of the Universidad Autónoma de

G. Bruno (✉) · N. Ruiz-López · C. S. de Castro
Faculty of Teacher Training and Education, Universidad Autónoma
de Madrid, Madrid, Spain
e-mail: gustavo.bruno@predoc.uam.es

N. Ruiz-López
e-mail: natalia.ruiz@uam.es

C. S. de Castro
e-mail: cesar.saenz@uam.es

© Springer International Publishing AG 2018
M. Jurdak and R. Vithal (eds.), *Sociopolitical Dimensions of Mathematics Education*, ICME-13 Monographs, https://doi.org/10.1007/978-3-319-72610-6_8

Madrid, aims to deepen the knowledge of primary and secondary schools located in "challenging" socio-economic contexts, addressing and approaching this research from a perspective of education for social justice.

From this perspective, education plays a fundamental role for the purpose of achieving a greater social justice and for minimizing the social inequalities. Education is one of the main ways of advancing towards more just and democratic societies. In GICE's research, we have assumed the approaches about social justice of Sen (2010) (redistribution), along with the ideas of Fraser (2008) (recognition and participation). From those principles, Murillo et al. (2011) elaborate an approach to the idea of education for social justice:

1. High quality education and just distribution: A pertinent and relevant education, with the same objectives for everyone involved; but also an education that devotes more efforts and resources to the people that, because of their origin, cultural background, native language, socio-economic context or capabilities, are more in need of resources and help.
2. Recognition and identity: An education that promotes the recognition, respect and appreciation of the individual, social and cultural differences.
3. Participation: An education that fosters and assures not only the learning, but also the participation of everyone in an environment of freedom and coexistence.

Thus, the triad of *Distribution*, *Recognition* and *Participation* constitutes the theoretical principle of the different research projects of GICE. This chapter draws on the research project "Schools in socio-economical challenging contexts: An approach from the perspective of Education for Social Justice.[1]" The project represents an update and prolongation of other GICE previous work, in which different schools, that identified themselves as committed to the social justice principles, were analyzed. An important part of this large-scale project included several case studies of teachers of mathematics and natural sciences, in six schools of Madrid in challenging contexts (both primary and secondary schools).

This chapter refers to one of those case studies and intends to determine the characteristics of social justice-related teaching. In order to accomplish this, we have simultaneously worked on two levels: first, an observational study of the lessons of a mathematics teacher in a secondary school in Madrid, on a particular subject that is part of the official mathematics curriculum; second and complementary to the observations, a biographical-narrative interview was conducted with the observed teacher. We consider as a point of departure the idea that, to be characterized as a teacher for social justice, we should be able to detect and identify, in the teacher, a vocation for transforming society, with the goal of reducing inequalities. Therefore, we analyze the data obtained from both methods contrasting and comparing what the teacher declares in the interview with his action in the classroom and in the school.

---

[1]Project Ref.: EDU2014-56,118-P.

2   Theoretical Framework

Attention to equity (Secada 1992), inclusion/exclusion (Knijnik 1993), social justice (Burton 2003; Gutstein 2003) or to democratic issues (Skovsmose and Valero 2008; Vithal 2003) goes back at least two decades on the agenda of mathematics education. The idea is that through mathematics we become "empowered" citizens, because mathematics is a central knowledge in the mapping of the social system, especially in a high-tech world with democratic ambitions. In this sense, achieving equity in mathematics education is equivalent to provide significant mathematics education to all students.

But Pais (2012) proposes a challenge to this ideal of "mathematics education for all". He notes that actually, not everyone will achieve success in learning mathematics. Thus, school mathematics becomes a powerful mechanism of selection and accreditation, a filtering technology for sorting peoples and capabilities for different social roles, and perhaps even an obstacle and a prerequisite to become a citizen. Straehler-Pohl (2015) considers that this selection mechanism may not even be about distributing different roles in society, but rather is about assigning adequate labels according to the success or failure in the school mathematics, in strong relation with socio-economic status. The purpose of mathematics education, as a social endeavor, seems not to necessarily be about the effective learning of mathematics, but about how mathematics education is a tool of social engineering, useful for the formatting of society, culture, and even the same subjectivities of people, be it the students (Andrade-Molina and Valero 2016) or the same teachers who become agents of that social engineering (Montecino and Valero 2016), for certain far reaching sociopolitical agendas. So, we have the unsettling possibility that marginalization and exclusion may not be merely residual problems of both the school and mathematics education, but its very nature and purpose as a social level.

Rasmussen (2010), in a similar sense, discusses the idea that mathematics education functions as a large-scale filtering and recruiting tool of mathematical talent, ensuring that the different social spaces and practices that need mathematical developments for its functioning (policymaking ambits, sciences, engineering and development of new technologies, economic agendas, etc.) can obtain fresh mathematical capabilities for continuing and refining its projects. Chartres (2008) considers that the learning of mathematics can result in empowerment and participation, or otherwise in the loss of power, marginalization and exclusion.

In mathematics education it frequently happens that, even if we acknowledge the influence mathematics has in the studies and process that allow people to reach power and an economic and political status, in the school it's usually presented as a "neutral science", without ideology, merely instrumental. Teachers are not usually conscious about the power of mathematics to favor social justice (empowering the future citizens with critical aptitude) or, on the contrary, to perpetuate inequalities (being mathematics a science that allows the elite to keep power and privileges) (Young 2008). The fact is that a large majority of young people today are subject to the teaching of mathematics during critical ages of their process of personal formation, that last almost the entire duration of compulsory education. It has an enormous influence in the formation and construction of their personality and identity, and this influence could be formative or deformative for the young person (Ernest 2010).

Thus, facing these problems, Vithal (2003) points to paradigms such as those of critical mathematics education, ethnomathematics, reflections on gender, race, class and equity, and the so called "people's mathematics" (in the context of post-Apartheid South Africa), as the core of a growing shift of interest on the sociopolitical and cultural dimensions in mathematics education. These paradigms constitute paths of reflection that try to find and show possibilities for implementing a more equitable mathematics education and to encourage citizens' participation. Authors who follow the paradigm of critical mathematics education propose the concept of Mathemacy (Skovsmose 1994; Chronaki 2010) as an essential element to the possibility of social and cultural emancipation. Lubienski (2002) believes that the key is to investigate how to train teachers capable of developing strategies to generate "math power" for all students, regardless of differences of social class, ethnicity or gender. Other authors (Forrest 1997; Frankenstein 2001, 2014; Osler 2007; Bateiha and Reeder 2014) have investigated the type of mathematical experiences that encourage students to develop a "critical numerical lens" to view and interpret the world. To do this, they propose the use of real-life contexts in an inter/transdisciplinary curriculum that relates mathematics to social or natural science.

Valero (2010) proposes the understanding of mathematics education as a network of social practices. That means going beyond the limits of different historically established paradigms of research in mathematics education (for example, the *didactic triad*), with the aim of providing better understandings and alternatives for the teaching and learning of mathematics, to face the socio-economic and cultural challenges of the present time. She also explores how this notion can envision new possible research paths, meanings and practices. Mathematics education is a complex and multi-layered phenomenon of social practice in which different actors, in realities beyond the limits of the classroom and the school, participate in different forms. Many of those actors have, unknowingly(?), a decisive role in mathematics education legal regulations, assessment methods, and even research agendas and results. So, her proposal implies also a reflection about the nature of the research field of mathematics education.

Our investigation shares with various authors (Bigelow et al. 2001) the necessity of transforming the teaching process to work for social justice. In this line, to characterize the teaching that works for social justice we have developed 12 indicators, as a synthesis and from the reflections related to the contributions of certain authors in this field (Banks 2004; Michelli and Keiser 2005; Cochran-Smith et al. 2009, 2010):

1. Commitment of the teacher to social justice (Banks 2004; Michelli and Keiser 2005).
2. High expectations of the teacher towards all the students (Cochran-Smith et al. 2009; Michelli and Keiser 2005).
3. Equitable and fair atmosphere promoted by the teacher and supported by the students.
4. Teaching strategies and activities that recognize and value the previous knowledge of the students and its inherent prejudices (Banks 2004; Cochran-Smith et al. 2010).
5. Cooperative work.
6. Active involvement of the students as part of a learning community (Cochran-Smith et al. 2009).
7. Varied teaching strategies that could be adapted to the different learning paces and characteristics of the students (Cochran-Smith et al. 2009, 2010).
8. Relations with other sciences and fields of study, and connections with the "real world" (Cochran-Smith et al. 2010).
9. Intellectual rigor, constructive criticism and appreciation of the students' challenging ideas (Cochran-Smith et al. 2009, 2010).
10. Questioning of the teacher inducing different ways of thinking (Cochran-Smith et al. 2009, 2010).
11. Atmosphere of mutual respect and appreciation.
12. Varied and alternative assessment methods (Cochran-Smith et al. 2009, 2010).

These 12 indicators provide a useful frame for the analysis of the results. But these are not specifically related to mathematics education, in light of the different perspectives we have mentioned before. So, a set of indicators directly linked to different sociopolitical and cultural perspectives of mathematics education could deepen and enrich our analysis of both the practices and the life story of the observed teacher. For this purpose, we propose also the following indicators:

A. The teacher recognizes and puts into practice the diversity of mathematical thoughts, and does not just prioritize "western European" mathematical thought [that is, the teacher shows awareness of the "mathematical enculturation" perspective developed by Bishop (1991) and the ethnomathematics point of view developed by D'Ambrosio (1985)].
B. The teacher takes into consideration the "informal" mathematics that is used in different social contexts (especially, the "informal" mathematics the students use, knowingly or not, outside the school, and in their other day-to-day activities).
C. The teacher acknowledges the linguistic capabilities of his students, attending to the diversity of languages and cultures, as a key element in the learning of mathematics (Planas and Civil 2007).
D. The teaching practice addresses sociopolitical and economic issues from a mathematics standpoint, and/or addresses the role of that mathematics plays in those issues (this idea is related to the key notion of the "formatting power" of

mathematics, from Skovsmose (1994); also, Frankenstein (1983) analyses the essential role of statistics for addressing these social issues).

E. The teacher does not consider Mathematics as a neutral, merely instrumental and separated knowledge (a "platonic" point of view), but rather as a knowledge constructed in the sociopolitical and historical reality of different communities, with different actors and interests involved (Skovsmose 1994; Ernest 2010; Valero 2010; Pais 2012).

F. The teacher shows an awareness of the complex and multilayered social phenomenon mathematics education is (in his practices, in his declared ideas, both or none).

3 Methodology

The school of the observed teacher is located in a disadvantaged area of Madrid with great ethnic, socio-cultural and religious diversity, and identifies itself (from management team and institutional ideal) as a school for social justice. The observed teacher agrees explicitly and emphatically with this paradigm.

The observations were carried out in eight 100-min sessions, which corresponded to the development (up to evaluation) of a subject matter in mathematics called "Functions". To organize the observation, we used the guide Reformed Teaching Observation Protocol and Social Justice items, developed by Pedulla et al. (2008), suitably adapted to the Spanish language and validated.

This guide combines the qualitative register of the activities developed in the classroom, with a checklist of different aspects of the development of each session. In concrete terms, the protocol is structured in five main sections:

(1) School data, identification of the teacher and the observer, time and date of the observation.

(2) Identification of the context of observation, including quantification and characterization of the people present in the classroom, and a description of the physical environment.

(3) Description of the observed class both globally and with a chronogram that temporalizes the activities of teachers and students in 5–10-min intervals.

(4) Checklist for the observation of the design and implementation of the teaching and learning processes. The observer punctuates, in a 0–4 scale, a list of indicators related to: the mathematical content developed in the class, class atmosphere (people interactions, teacher-student relationships), and social justice issues.

(5) Interviewing the teacher after the observation with questions such as: "Was today a typical class day? Why, why not? Did the lesson go as planned? Do you sometimes make changes in the curriculum? Can you describe them?", etc.

Besides these observations, the research included a biographical narrative interview. This was a recollection and a reflection of episodes of the biography of the teacher, in a framework of an open exchange (introspection and dialogue) allowing the insight of particular circumstances of his life and the active listening of the interviewer, who then develops a final report with the information and impressions obtained (Ruiz-López et al. 2015). For this reason, a set of questions about the teacher's childhood were posed, his first contacts in education, social engagement experiences, etc., to reconstruct part of the life story of the interviewee.

We consider that this biographical-narrative interview is a valuable complementary tool of the observation guide, for achieving a better understanding of the class observations, and thus, for achieving a more complete picture, a more detailed portrait, of the observed teacher. Indeed, the understanding of the different actions, approaches, methodologies, evaluation methods, attitudes and decisions that constitute the day-to-day activities of this mathematics teacher can be related (as we well see in the analysis) to the previous life experience, professional training, and personal ideals of the teacher. All this information emerges from the biographical-narrative interview.

4 Case Study

The observed student group was in the second year of compulsory secondary education (2nd ESO). The class had 50 students in total, 24 girls and 26 boys. There were 17 nationalities in the group: Spanish, Moroccan, Ecuadorian, Peruvian, Dominican, Paraguayan, Bolivian, Uruguayan, Chilean, Swiss, Egyptian, Romanian, Palestinian, Syrian, Philippine, Chinese, Algerian. The Spanish nationality was the most numerous (17); but the Muslim students were a majority of 20 in total. The group included three students with diagnosed learning difficulties, for whom there were dedicated tutoring periods (both inside and outside the class) in which the observed teacher also participated with other colleagues of the same school. The expected age for 2nd ESO is 13–14 years old. However, the observed age of the students was in the 13–16 year old range.

The total group of students were organized for cooperative work. They were divided into 11 teams with four students each and two more groups with three students, and each team had its own desk. These teams were formed by the teachers according to both academic and social integration criteria. The teachers reorganized the teams every other trimester.

The mathematics lessons were developed in what it is called "shared-space", together with the natural science subject. That is, the class had 50 students with sections A and B joined, and had three teachers for the shared space mathematics-natural science (two natural science teachers and the mathematics teacher). Lessons for both subjects were performed at different moments by their respective teachers, and during the observations the contents of mathematics and natural science were being worked separately (no content was addressed in a unified approach from both subjects). Nonetheless, the three teachers were involved in all the lessons.

The observed teacher was young and recently graduated; although with previous teaching experience in contexts other than the school. He had only one year of experience as a teacher of this school, and the other two natural science teachers were his senior, more experienced partners. However, the relationship between them was completely egalitarian in terms of decisions and agreements of the strategies to follow. This was further verified by the observer after reunions and conversations with the teachers, in their lounges and during breaks.

In the observed lessons, the mathematics topic developed by the teacher was "Functions". At the same time, in the natural science classes the content was "Kinematics". The mathematics teacher started the subject matter with a formal concept of "function", and afterwards he followed with explanations on the different ways of expression and representation of functions (using words, tables or lists, graphics or formulas), and then concluded with the topic "linear functions and affine functions". The usual procedure of explanation was "definition-example—questions-exercises/problems", in an expositive way. In very specifics occasions, the explanations were accompanied by examples that somehow tried to connect the contents with the background of the students. In parallel, the natural science teachers started with the basic concepts of Kinematics (trajectory, displacement, distance, time, velocity), and in later lessons they put the focus in the classical "Uniform Rectilinear Motion" paradigm.

Some days, the group of students were divided into three subgroups, one for each teacher. One of those groups remained in the large classroom with one teacher, solving exercises or preparing for an exam, theoretically still in the structure of cooperative work. The second group went with another teacher to a smaller sub-classroom devoted to more dedicated tutoring or to individual explanation, and there the usual structure of cooperative work was essentially dissolved (i.e., the class became a "normal" class). The third group went with the third teacher (one of the natural science teachers, during the time of the observation) to work in the science laboratory. The observer didn't have the opportunity to attend one of those classes in the laboratory, prioritizing always the observation of the mathematics teacher.

5 Results

The results can be described as follows: the teacher explicitly declares in the interview his ideals linked to education for social justice, and he develops his practice to promote these ideals in many ways and with different resources. But this is done in a way "extrinsic" to the disciplinary content, a way we could consider to be (almost entirely) outside of mathematics. In other words, the "value neutral-instrumental" perspective on mathematics was prevalent. Nonetheless, his observed professional activities are consistent with what he stated in the interview (Sáenz de Castro et al. 2015).

We analyze the information obtained from both the observations and the biographic-narrative interview in the light of the indicators on education for social justice, and those more specific to mathematics education from a sociopolitical perspective, proposed in the theoretical framework.

In the interview the teacher said that he came from a family of educators, that he has an affinity for mathematics and that he had an early affinity for education, with the perspective that a teacher has the chance to work for a better world and to be an agent for social change. His teaching methods and procedures met many of the criteria we cited earlier about what a quality mathematics education for social justice could be. He also showed a special commitment for working in a school in a difficult context and with a population with a significant part of students with complex backgrounds (in terms of the usual political correctness, this means immigrants, low-income families, etc.). We observed indicator 1 reflected here.

The teacher promoted a fair and just environment for every student in the class, an atmosphere of mutual respect between students and teachers. There were not any conflicts during the observations, notwithstanding the great diversity of the people involved, nor any situation that could interrupt the normal development of the lessons. It was observed that there was no preferential treatment or priority for the most effective (i.e., "higher scoring") students. The teacher had appreciation and high expectations for all his students, he knew them very well, their problems and contexts. There were no comparisons or labels. The teacher acknowledged the most effective students and was always close to those students most needing help, recognizing their achievements and progress. The adjectives used by the teacher always highlighted the positive features, such as "she's very smart", "he puts a lot of effort", "when he concentrates he is very fast", and he never applied negative adjectives. His expectations were that everybody could learn and pass the course. We observed indicators 2, 3 and 11 reflected here.

Both the teacher and the school had the policy of attention to the diversity of contexts, backgrounds and learning paces. There was specialized dedicated tutoring for students with learning difficulties (both temporary for personal reasons, or for more permanent conditions), imparted by the same teachers of the observed class and others. We observed indicator 7 reflected here. We note, however, that for this teacher and classes, we did not have the opportunity to obtain more knowledge about how, or with which criteria, these "learning difficulties" diagnosis was assigned to some students.

Other aspects to highlight are the varied and alternative evaluation methods. The teacher tried to soften the supremacy of the written exam, composing the final score from the results of diverse activities, recognizing the responsibility of doing the homework and the positive attitudes, with a day-to-day observation, and notably, a consideration for the particular situations of the students (in the cognitive order, but also in the familiar, social, cultural, etc.). The traditional written individual exam that the teacher conducted at the end of the subject matter represented only a 40% of the final score. The teacher, thus, had managed to effectively lighten the weight of the traditional written exam in the final score, with a careful daily work and specific internal rules for the mathematics-natural science space (agreed with the other teachers and perfectly known and maintained by the students). Indicator 12 is clearly reflected here.

However, regarding the disciplinary content, we noticed first a far too theoretical approach to the "Functions" topic, "from the general to the particular". The strategy was expositive-descriptive, following the scheme "definition/proposition-example-exercise", reinforcing and repeating when necessary, and the attitude of the students was primarily passive-receptive. Most of the students' interventions during explanations were in the form of questions or doubts about the contents. There were few opportunities for the students to "challenge" the proposed concepts (and for the teacher to appreciate it) or to be actively involved in the approach to the subject, due to the high degree of abstraction, and the examples being artificial or artificially "realistic". Connections with the students' daily life in and outside the school were given as specific or sporadic occurrences. No attempt was made to take advantage of the potential richness of the cultural, ethnic and linguistic diversity of the students. This analysis is related mostly to indicators 4, 6 and 8, and indicators A and C on the sociopolitical dimensions of mathematics education.

On a few occasions, the teacher developed certain approaches to the content that directly addressed the life of the students outside the school, and these were more contextualized in their day-to-day experiences. With the concept of function, for instance, he tried to exemplify the very abstract concept of "relation between variables/sets" with the functional relation between the players of "La Roja", the Spanish football team, and their respective squad numbers. That example was perfectly understood by the students, although we cannot know if they indeed connected that "relation between players and numbers" with the language of "variables x, y", or if this example was finally useful for them for attaining the desired concepts and succeeding in the resolution of exercises.

On another occasion, this time about the concept of "velocity", the teacher made an interesting intervention during the explanation being given by one of the teachers of natural science. He put in the computer connected to the projector, for everyone to see, an internet meme with one of the characters of the series "Futurama". In this meme, the character says "my girlfriend told me she needs time and space... she must want to calculate velocity". Again, the meme, and the intervention connected to some aspect of the urban culture of the students, was well received by them, even with laughter. However, again, we do not really know how much the students finally connected this example with the concepts and formulas proposed during the lessons. Both examples were isolated occasions in the usual classrooms procedures, but they showed at least an awareness by the teacher, in the form of intuitions and witticisms, of possible ways of connecting the disciplinary contents with the students' urban culture. As we remark later, the teacher labelled these situations as "bright ideas". The analysis in this paragraph and the previous one can be related to indicators 8 and B on the sociopolitical dimensions of mathematics education.

It must also be noted that in the observed sessions, we cannot say that the content and the activities (exercises, problems) were planned from the perspective of cooperative work. It was observed, in only one class, an activity in which students were required to solve certain problems working together, and then each student wrote the results in his or her own notebook. But these problems did not demand per se any cooperative work, and they could have been perfectly solved by students individually. Thus, the subject content, exercises and problems were not really designed for cooperative work. Curiously, the students showed a marked tendency to solve the mathematics problems thinking together in small groups, or working in couples.

For this reason, we believe that cooperative work was perhaps more a goal of the school in general and not a precise objective planned for the space of mathematics-natural science. Indicator 5 is related to this analysis.

The structure of shared-space with natural science, an attempt for an interdisciplinary approach to mathematics and natural science, showed several advantages (such as linking the concept of linear and affine function with the principles of the kinematics), but most importantly, it also implied an instrumental view of mathematics in relation to the natural science. The subject-matter of mathematics and natural science were otherwise virtually separated in terms of development and evaluation. This can be related to indicators 8 and E.

6 Discussion

We give form to the discussion of the results also in the light of the considerations and the indicators set out in the theoretical framework.

As a first approach, we could consider that the remarks of the previous section suggest many inconsistencies between what the teacher says about his ideals on social justice and his actions in the classroom. However, from the interview we know that during the first years of his undergraduate and graduate studies he suffered some disappointment with mathematics. He perceived it as "cold and dehumanized", far away from the "real world" (he considered that higher education as a whole had this character), not allowing him to help other people and to fill his expectations of "social commitment". He tried to compensate for these shortcomings by participating in solidarity social organizations, and he considers that he improved "exponentially" as a teacher with them, reaffirming his vocation as an agent of social change. However, it all happened in a way "external" to mathematics itself and its contents. He thus never approached, at least during his training years, mathematics or mathematics education from a sociopolitical, cultural or economic perspective. Thus, in no way an idea like "mathematics education as a network of social practices" could have been reflected or addressed during his formation as a teacher. After his training, he was left only with his pure vocation and intuitions about how to develop a significant and socially compromised mathematics education, if he really wants to do it and become "an agent for social change" instead of an agent of social reproduction. The teacher seems to be intuitively at odds with the expected or desired cultural role of a mathematics teacher, in revolt with the molding of his subjectivity (Montecino and Valero 2016). Indicators E and F shed light on the previous life experience of the teacher, to better understand his actions in the school and classroom.

Regarding teacher training, he had a very critical posture, considering that many of the people involved in the education of future teachers had no experience whatsoever in primary or secondary education (a statement that should be analyzed in the context of the Spanish system of mathematics teacher training). His literal expression was "this person has not seen a teenager in 20 years", pointing to the disconnection between the realities of teacher training and teacher practice. So he perceived that his training and formal education, and the mathematics he learned during this process, in no way allowed him to attain his aspirations of social commitment. His views on mathematics education research and its results were critical too, considering, in his experience, that those results seldom reach the "real" classroom or are useful for implementing and improving the day-to-day teaching practice and the learning of the students. The teacher thus shows awareness on the issues about mathematics, mathematics education and mathematics education research set out in indicators D, E and F. Also, we observe a clear critical posture on the field of mathematics education research. It is important to note that this reflection of the teacher could not have been obtained without the biographic-narrative interview.

But he also declared in the interview, that in terms of values, and in terms of social significance, mathematics instills in the students the idea of perseverance, self-discipline, something that teaches them to face up to problems, to overcome difficulties, and thus they can supposedly transfer those qualities to problems and difficulties in their daily lives (present or future). This would seem to mean, at a first glance, that the teacher agrees with the disciplinary role and the molding of subjectivities that some authors describe (see Andrade-Molina and Valero 2016, p. 253, citing many other authors) for mathematics education as a social endeavor, a "technology of the self".

But this could be an incomplete picture of the teacher. Because, also in the interview, he talked about some projects he had organized in another class (not the one observed) around statistics, probability and gambling addiction. He stated that he organized a dice game with his students, and after a while, he proposed to them a reflection about how, in gambling, a person would most likely lose everything at the end. He also commented in the interview that he knew works in ethnomathematics. Furthermore, the teacher showed the observer a project he had developed for the third course of secondary education, to propose to his students the exploration of geometry in some of the Islamic monuments of Spain (*La Alhambra de Granada* and *La Mezquita de Córdoba*). He was addressing the Muslim background of many of his students, and also exploring a mathematics distinct not only from the usual school practice but also from the most mainstream "western modern" mathematics. Thus, we connect again with indicators A, B, D and E.

At the time of the interview, he believed that addressing sociopolitical significance of mathematics in mathematics education is indeed a possibility. But he considers that the sociopolitical and cultural significance of mathematics and mathematics education is somehow "hidden". The teacher in fact acknowledges that it is possible to approach mathematics from the ideals of social justice and from sociopolitical considerations, but believes that this is very difficult, something that he still does not see how it can be done on a continuous basis but rather only sporadically and in moments of inspiration. He considers that the teacher cannot always have "bright ideas". The curriculum, calendar and assessment demands, and in general, the structure of the school and educative system, were also considered by him a strong influencing factor in the teaching possibilities. Again, we observe a greater awareness by the teacher of the complex sociopolitical phenomenon mathematics education is (indicators E and F) that what could have been obtained at first from the class observations alone.

7 Conclusion and Implications

As a first conclusion, we note the importance of developing a different approach to mathematical contents and its teaching strategies when we look for social justice-oriented teaching. The strategy followed during our class observations can generate, even with the clear aim of teaching for social justice, the opposite effect, and convey the opposite values to what is intended. Indeed, the strategy followed by the teacher in approaching the content "from the general to the particular", starting with abstract concepts and definitions totally unknown by the students, tends to generate a context in which students are first and foremost "recipients of information" and not "actors" in the process.

This strategy strongly limits possibilities such as the active involvement of the student group, the exploration of the richness of cultural and ethnical diversity, and the cooperative work. The examples, exercises and problems studied during the lessons were mostly decontextualized, artificial (disguised as "real-life" problems); and in some cases, can induce in the students a perspective towards mathematics as a "neutral science", free of ideology, purely instrumental, with no relation whatsoever with the fabric of the social system in which we all live (Skovsmose 1994). Taking into consideration the three principles of the GICE research project, we could say that the approach to the mathematical content severely limited the possibilities for the students' participation.

But, as we have explained before, the diverse "exceptions" observed in the classes, or obtained from the interview, creates a contrast with the previous conclusion, so we cannot really consider that the teachers' view on mathematics and mathematics education was purely instrumental and value-neutral. One of the explanations we could offer about his teaching strategies was that he was not the only teacher in the classroom, and his partners were more experienced. The content to teach was not entirely decided by him, and the instrumental point of view on mathematics could have been implicit in the relation established with natural science. The teacher's own strategies and "creativity", shown in other instances, could have been conditioned by the very format of the "shared-space" and the curriculum. This interdisciplinary approach, however, showed some interesting advantages, as we said before.

So, the following questions unavoidably emerge: how far, if possible, we can go with our aspirations for a genuine social justice oriented teaching, especially on mathematics education, in the present format of the curriculum, the school rules and culture, class context and assessment procedures? In other terms: how much can be done in the apparently noble aspiration of social justice with an abstract, separated and procedural content as the one the teacher was trying to teach "Functions"? From some of the authors we have cited before, we should ask again, if the marginalization and exclusion (in the form of a social technology of selection and accreditation) are not merely a residual effect of mathematics education, but its very nature.

A more relevant explanation and conclusion from the above would be, however, the need to develop and propose a teacher training specifically oriented to social justice education, at least for those teachers who want to develop a mathematics education for social justice. The investigative literature is coherent with that; as Cochran-Smith says (2009), only if teacher training is seriously approached with a view towards social justice, will it be possible that gets in the classrooms. The complex and global social changes and the diversity of scholar contexts imply the necessity of constant and better studies that deepen the specific educational elements and factors for teachers to achieve (in the face of such diversity) the formation of socially and culturally capable, ethical and competent people. Those elements, conditions and challenges must be part of the models of teacher training, and also specifically for mathematics teachers.

As we explained in the beginning of this chapter, the GICE's research program has the objective of developing different case studies of mathematics teachers in the context of education for social justice. Of the different teachers observed and interviewed, the one studied in this chapter was perhaps the most assertively convinced about the principles of social justice. He also clearly articulated his ideas, and effectively worked in his classes trying to achieve his goals, adjusting his strategies and participating in different ways in the general objectives of the school. We did not find such conceptual clarity about social justice, and coherence between declared ideas and effective actions, in any of the other teachers studied. But his training as a mathematics teacher has had an impact on the way he focused on the disciplinary contents and the apparent separation he established between mathematics and the social justice-oriented values. The teacher is convinced of his ideals, but his mathematical training had almost made him doubt his vocation and even today he does not see clearly how mathematics education can contribute to his ideals of social justice.

The instrumental, value-neutral, aseptic perception on mathematics and mathematics education is at present in the teacher training and in higher education as it is in the school. Thus, this chapter tries to contribute with some ideas and criteria to teach and act for social justice through mathematics' curriculum and teacher training. The set of indicators identified in the investigative literature, both those referred to education for social justice in a general sense, and those specifically concerned with mathematics education, was found to be a useful means to analyze the results of the observations. We believe it could constitute the nucleus of an effective formative program for social justice-oriented teachers, and also, as Valero (2010) suggests for the perspective of "mathematics education as a network of social practices", a set of notions with a rich potentiality to open new insights and research paths. Work to operationalize these indicators in the mathematics learning and teaching process still must be done.

We should note, moreover, some limitations in our study. The main restriction is that our class observations lasted just one month and we covered only one subject matter. Therefore, we are missing data about the strategy the teacher uses in other lessons at the same school and if he takes advantage of the social and cultural context in different situations (the teacher had other groups of students in the same school). Besides, he declared that many strategies were needed to meet the curriculum demands, assessments and scores (the observation was realized during April–May); and he worked in a space shared (both physically and in the sense of the curricular contents) with other teachers, so his possibilities for implementing different strategies were also limited. Concerning this, the observer-interviewer had the chance to know, from the interview and from diverse conversations, about a different didactical approach implemented by the same teacher in another course of the school, as we mentioned before.

During the presentation of a previous version of this work in the Topic Study Group (TSG) 34 of the ICME-13 (in Hamburg, Germany, 2016), the questions of some of the participants also helped us to gain insights in the limitations and precautions we should have with this study.

One of the questions was about if we have told and informed the teacher about the many conceptual frameworks (mathematics education for social justice, the value-neutral and instrumental approach, the interweaving of mathematics with power structures in society, etc.) with which we were evaluating and apparently criticizing his professional work. The answer was: of course, no, and in fact this question helped us realize how critical the teacher training and his life story is, and how significant the proximity of the researcher with the "real school" is. We also remarked that we considered him a very good teacher, and in this new version we have cited more of the notable interventions and activities in which the teacher showed a notorious (in comparison with other case studies), if intuitive, awareness about many of the issues we are here discussing. But, as Pais (2016) explores in his work, we should be cautious and ready for self-inspection of our roles as researchers. Little we would help the endeavors of the professionals entangled in this education for social justice issue if we remain merely critical and distanced from the school practice once our research interests shift attention. Even more, we could end up being complicit with the dominant practice of exclusion, or we could simply end up replacing a dominant discourse with a different one, purely derived from a self-contained and disconnected world of research.

The second helpful question, in a similar line with the previous one, was if we really believed that social justice could reach the classroom. In our answer, we acknowledged that we should not understand social justice as a platonic entity that can "walk inside a mathematics classroom", because, again that would simply mean that we are replacing a dominant discourse with a new one, a discourse formulated in the sphere of research away from the reality in which we pretend to participate. We must not be tempted to announce social justice as a new grandiose ideal; our work is to be interpreted as cautious approach to certain educational struggles and challenges concerning socio-economic and cultural inequalities. An approach in which we focus the attention on the sociopolitical role of mathematics and mathematics education, trying to understand and participate in those struggles as much as to assess them.

Also, it is important to note that at the present, at the level of official discourse, the Spanish education system is in fact in a moment of transition for the implementing of the most dominant Competences paradigm, and to make reforms and take measures in order to improve the results in international assessments such as PISA (Sáenz and García 2015; Álvarez 2016). But this shift of paradigm has its own deal of problems and difficulties (Sáenz and García 2015), so there is at the present an opening of possibilities and paths of educational change, before the dominant paradigm takes its full grip on Spanish education. In this breach, we are trying to address the educational initiatives concerned with questions of social justice that arise as a natural response to the social challenges of inequality, diversity, immigration, unemployment, etc. Our intention with the more general GICE project and with these special case studies is to gain an insight and participate in a social tendency towards the opening of those educational possibilities.

It will also be important, for a future research study, to consider an approach to the students' experiences and points of view, trying to find out their beliefs and values about education, mathematics, teachers and issues of social justice. This study would bring relevant data to evaluate the efficacy of the imparted teaching on the aim to supply social values to the students.

References

Álvarez, P. (December 3, 2016). Informe PISA. La educación española se estanca en ciencias y matemáticas y mejora levemente en lectura. *El País*. Recovered from: https://politica.elpais.com/politica/2016/12/05/actualidad/1480950645_168779.html. Accessed July 28, 2017.

Andrade-Molina, M., & Valero, P. (2016). The effects of school geometry in the shaping of a desired child. In H. Straehler-Pohl, N. Bohlmann, & A. Pais (Eds.), *The disorder of mathematics education—Challenging the socio political dimensions of research* (pp. 251–270). New York: Springer.

Banks, J. A. (2004). Teaching for social justice, diversity, and citizenship in a global world. *The Educational Forum, 68*(4), 296–305. London: Taylor & Francis Group.

Bateiha, S., & Reeder, S. (2014). Transforming elementary preservice teachers' mathematical knowledge for and through social understanding. *Revista Internacional de Educación para la Justicia Social (RIEJS), 3*(1), 71–86.

Bigelow, B., Harvey, B., Karp, S., & Miller, L. (Eds.). (2001). *Rethinking our classrooms: Teaching for equity and justice*. Milwaukee, WI: Rethinking Schools Ltd.

Bishop, A. (1991). *Mathematical enculturation—A cultural perspective on mathematics education*. Dordrecht: Kluwer Academic Publishers.

Burton, L. (2003). *Which way social justice in mathematics education? International perspectives on mathematics education*. Santa Barbara: ABC-CLIO.

Chartres, M. (2008). Are my students in engaged in critical mathematics education? In J. F. Matos, P. Valero, & K. Yasukawa (Eds.), *Proceedings of the Fifth International Mathematics Education and Society Conference* (pp. 23–45). Lisbon: Centro de Investigação em Educação.

Chronaki, A. (2010). Revisiting Mathemacy: A process-reading of critical mathematics education. In H. Alrø, O. Ravn, & P. Valero (Eds.), *Critical mathematics education: Past, present and future* (pp. 31–49). Rotterdam: Sense Publishers.

Cochran-Smith, M. (2009). Toward a theory of teacher education for social justice. In M. Fullan, A. Hargreaves, D. Hopkins, & A. Lieberman (Eds.), *The international handbook of educational change* (pp. 916–951). New York: Springer.

Cochran-Smith, M., Gleeson, A. M., & Mitchell, K. (2010). Teacher education for social justice: What's pupil learning got to do with it? *Berkeley Review of Education, 1*(1). Resource document http://escholarship.org/uc/item/35v7b2rv#page-1. Accessed: July 31, 2017.

Cochran-Smith, M., Shakman, K., Jong, C., Terrell, D., Barnatt, J., & McQuillan, P. (2009). Good and just teaching: The case for social justice in teacher education. *American Journal of Education, 115*(3), 1–48.

D'Ambrosio, U. (1985). Ethnomathematics and its place in the history and pedagogy of mathematics. *For the Learning of Mathematics, 5*(1), 44–48.

Ernest, P. (2010). The scope and limits of critical mathematics education. In H. Alrø, O. Ravn, & P. Valero (Eds.), *Critical mathematics education: Past, present and future* (pp. 65–87). Rotterdam: Sense Publishers.

Forrest, M. (1997). Literacy and numeracy. *ALEA, Language in Mathematics Newsletter, 8*, 1–10.

Frankenstein, M. (1983). Critical mathematics education: an application of Paulo Freire's epistemology. *Journal of Education, 165*(4), 315–339.

Frankenstein, M. (2001). Reading the world with math: Goals for a critical mathematical literacy curriculum in mathematics. *AAVV, Mathematics: Shaping Australia Conference Proceedings 18th Biennial Conference of the Australian Association of Mathematics* (pp. 53–64). Adelaide, SA: AAMT.

Frankenstein, M. (2014). Which measures count for the public interest? *Revista Internacional de Educación para la Justicia Social (RIEJS), 3*(1), 133–156.

Fraser, N. (2008). *Escalas de justicia*. Barcelona: Herder.

Gutstein, E. (2003). Teaching and learning mathematics for social justice in an urban, latino school. *Journal for Research in Mathematics Education, 23*(1), 37–73.

Knijnik, G. (1993). An ethnomathematical approach in mathematical education: A matter of political power. *For the Learning of Mathematics, 13*(2), 23–25.

Lubienski, S. (2002). Research, reform and equity in U.S. mathematic education. *Mathematical Thinking and Learning, 4*(2–3), 103–125.

Michelli, N., & Keiser, D. (Eds.). (2005). *Teacher education for democracy and social justice*. New York: Routledge/Taylor & Francis.

Montecino, A., & Valero, P. (2016). Mathematics teachers as products and agents: To be and not to be. That's the point!. In H. Straehler-Pohl, N. Bohlmann, & A. Pais (Eds.), *The disorder of mathematics education—Challenging the socio political dimensions of research* (pp. 135–152). New York: Springer.

Murillo, F. J., Román, M., & Hernández-Castilla, R. (2011). Evaluación educativa para la justicia social. *Revista Iberoamericana de Evaluación Educativa, 4*(1), 7–23.

Osler, J. (2007). *A guide for integrating issues of social and economic justice into mathematics curriculum*. Resource document http://www.radicalmath.org/docs. Accessed: December 15, 2016.

Pais, A. (2012). A critical approach to equity. In O. Skovmose & B. Greer (Eds.), *Opening the cage. Critique and politics of mathematics education* (pp. 49–92). Rotterdam: Sense Publishers.

Pais, A. (2016). The narcissism of mathematics education. In H. Straehler-Pohl, N. Bohlmann, & A. Pais (Eds.), *The disorder of mathematics education—Challenging the socio political dimensions of research* (pp. 53–63). New York: Springer.

Pedulla, J., Mitescu, E., Jong, C., & Cannady, M. (2008). *Observing teaching for social justice for teachers from two pathways.* Resource document. Annual Meeting of the American Educational Research Association. https://www.researchgate.net/publication/267233961_Observing_teaching_for_social_justice_for_teachers_from_two_pathways. Accessed: January 10, 2017.

Planas, N., & Civil, M. (2007). Reconstrucción de creencias, prácticas e identidades en torno a la educación matemática de alumnos inmigrantes. In J. Giménez, J. Díez-Palomar, & M. Civil (Eds.), *Educación matemática y exclusión* (pp. 131–146). Barcelona: Graó.

Rasmussen, P. (2010). The critical perspective on education and on mathematics education. In H. Alrø, O. Ravn, & P. Valero (Eds.), *Critical mathematics education: Past, present and future* (pp. 161–169). Rotterdam: Sense Publishers.

Ruiz-López, N., Atrio Cerezo, S., Bosch Betancor, J., & Bruno, G. (2015). Características biográficas del docente de matemáticas para la justicia social en educación secundaria. In P. Scott & A. Ruiz (Eds.), *Educación matemática en las Américas: 2015* (Vol. 5, pp. 56–65). Etnomatemática y sociología República Dominicana: CIAEM.

Sáenz, C., & García, X. (2015). *Matemáticas: Placer, poder, a veces dolor. Una mirada crítica sobre la matemática y su enseñanza.* Madrid: UAM Ediciones.

Sáenz de Castro, C., Bruno, G., Ruiz-López, N., & Atrio Cerezo, S. (2015). Estudio observacional sobre la docencia en matemáticas para la justicia social. In P. Scott & A. Ruiz (Eds.), *Educación matemática en las Américas: 2015* (Vol. 5, pp. 126–137). Etnomatemática y Sociología República Dominicana: CIAEM.

Secada, W. (1992). Race, ethnicity, social class, language, and achievement in mathematics. In D. A. Grouws (Ed.), *Handbook of research on mathematics teaching and learning* (pp. 623–660). New York: Macmillan.

Sen, A. K. (2010). *La idea de la Justicia.* Madrid: Taurus.

Skovsmose, O. (1994). *Towards a philosophy of critical mathematics education.* Dordrecht: Kluwer Publishers.

Skovsmose, O., & Valero, P. (2008). Democratic access to powerful mathematical ideas. In L. D. English (Ed.), *Handbook of international research in mathematics education* (pp. 415–438). New York: Routledge.

Straehler-Pohl, H. (2015). Mathematics, practicality and social segregation. Effects of an overtly stratifying school system. *Revista Internacional de Educación para la Justicia Social (RIEJS), 3*(1), 55–70.

Valero, P. (2010). Mathematics education as a network of social practices. In V. Durand-Guerier, S. Soury-Lavergne, & F. Arzarello (Eds.), *Proceedings of the sixth congress of the European society for research in mathematics education* (pp. LIV-LXXX). Lyon: Institute National de Recherche Pedagogique.

Vithal, R. (2003). *In search of a pedagogy of conflict and dialogue for mathematics education.* Dordrecht: Kluwer Academic Publishers.

Young, I. M. (2008). From constructivism to realism in the sociology of the curriculum. *Review of Research in Education, 32,* 1–28.

Outcome of the Market: The Outdated Mathematics Teacher

Alex Montecino

Abstract This chapter seeks, on the one hand, to illustrate the configuration of a mathematics teacher who is always considered to be outdated, and on the other hand, to discuss the circulation of a promise of salvation embodied in the discourses of permanent training. This chapter aims to contribute to the problematization of "what the mathematics teacher must be" and power effects in the fabrication of mathematics teachers' subjectivities. A Foucault-inspired discourse analysis is deployed in order to unpack naturalized truths, as well as forces that govern and control teachers. It argues that current research on mathematics teacher frames teachers within a narrative that is characterized by a continuous enunciation of new repertoire of techniques, practices and knowledge that the teacher should have, to become successful. New social demands and interests are conducting teachers into investing more and more in themselves as the only way to improve and to not become outdated in order to stay in the system.

Keywords Mathematics teacher · Discourse analysis · Power effects
Permanent training

1 Introduction

Nowadays, everybody has something new to say about mathematics teachers and their roles, education, quality, responsibilities and performances, within which performances, seems to establish the idea that the mathematics teacher always has to improve. Studies about the mathematics teacher, for example, those produced by researchers of mathematics teachers and those that support various Organisation for Economic Co-operation and Development (OECD) reports, constantly (re)produce discourses which characterize the effective and successful teacher (see Jacob et al. 2017; OECD 2012). They circulate ideas of how mathematics teachers must be, the

A. Montecino (✉)
Aalborg University, Aalborg, Denmark
e-mail: alexm@plan.aau.dk

© Springer International Publishing AG 2018   151
M. Jurdak and R. Vithal (eds.), *Sociopolitical Dimensions of Mathematics Education*, ICME-13 Monographs, https://doi.org/10.1007/978-3-319-72610-6_9

desired practices, knowledge and outcomes that mathematics teachers should have, as well as what is needed to develop or improve such teachers. They provide a way of thinking and understanding the mathematics teacher, through the (re)production of a network of discourses that are constantly reconfigured in response to social interests, problems, changes and demands.

These studies respond to the social concerns of getting better outcomes and overcoming the problem of failure in school mathematics, through the articulation of new practices, methods or techniques, supported by scientific knowledge. The desired mathematics teacher is articulated through the idea that mathematics is important for the development of persons, society and economy. Nowadays, it is considered that higher education achievement improves opportunities in the labor market and earnings expectations, thus benefiting the individual and their social well-being (OECD 2014a). Likewise, UNESCO (2007) has acknowledged that "mathematics education is a key to increasing the post-school and citizenship opportunities of young people" (p. 6). It is therefore seen as vital to improve the quality of teaching and learning of mathematics, as well as the quality of the people responsible for teaching mathematical knowledge to new generations—mathematics teachers.

Gutierrez (2013) asserted that research on the mathematics teacher seems to have the aim of developing and promoting successful mathematics learning experiences for students. Through this aim, research is set as a means to encourage the improvement of all aspects of the teacher that are considered deficient, producing statements that are established and acted upon as naturalized truths—a contingent effect of relations of force (Ribeiro 2011). The mathematics teacher is configured as a product and agent (Montecino and Valero 2017), in other words, an object of policy that is configured for consuming and promoting valuable knowledge for society: the mathematical knowledge. Montecino and Valero (2017) show how international agencies give evidence of how it is possible to intervene so that the teacher becomes the best version of such product and agent. Moreover, mathematics teacher research constitutes an idea of the desired teacher and what characterizes one, establishing as truth that mathematics teachers must have knowledge and master a repertoire of techniques for their actions and performance, as well as possess personal attributes in accordance with their practices, but teachers are in a social context and have deficits that present obstacles to their effectiveness (Montecino 2017). The competitiveness and effectiveness of the mathematics teacher are constituted as the main focus for promoting ways of understanding and thinking about the teacher (op. cit.). This reduces mathematics teachers to the effectiveness and competitiveness that they have in the educational system, with respect to some standards or desires. Mathematics teachers become professionals, governed and governing themselves by a neoliberal rationality, in which expert knowledge and capitalist logic of consumption influence the becoming of teachers and their productiveness, effectiveness and competitiveness (op. cit.).

The circulating discourses show, on the one hand, how the mathematics teacher has to face new challenges and requirements, through practices, repertoires of techniques and knowledge considered successful and valuable. For example, society requires that students develop an engagement with mathematics. A study by Skilling et al. (2016) explores secondary mathematics teachers' perceptions of student engagement in mathematics, concluding that it is "important for teachers to assess their personal beliefs about student engagement and consider how their practices in mathematics classrooms may or may not be supportive of students' mathematical engagement and learning" (p. 564). That is, through the development of tools and knowledge in the form of their capacity to reflect on their beliefs and practices, the mathematics teacher will become effective. On the other hand, in their constant search for the effective and competent mathematics teacher, studies reveal how some practices, repertoires of techniques and knowledge become obsolete or less effective, circulating the idea that the mathematics teachers have to be in a permanent process of training and improvement. This process is constantly changing as a function of social interests, demands and ideas of what is desirable. It is established as feared or undesirable for teachers not to reach the expected outcomes or levels of achievement, as well as for them to fail to keep up with rapid social changes and novel demands.

Thus, this chapter seeks to show how a mathematics teacher is always configured to be outdated. It is problematizing the constituting of the mathematics teacher, which is framed in a production strategy based on what is desired and on market logic, as well as the promise of salvation that responds to the fear that mathematics teachers may become ineffective or incompetent. Hence, it will propose that consumerism has become configured as the only method that mathematics teachers have for improving and not becoming outdated. In other words, the success of mathematics teachers depends on their investment in and consumption of permanent training. The contention of the chapter is that the mathematics teacher has to consume more and more training to stay in the system, having the constant risk of becoming an inefficient, not useful and valueless teacher. It also contends that the mathematics teacher cannot just be understood as a subject that has the job of teaching mathematics; the mathematics teacher cannot be reduced to a specific set of knowledge and practices useful for teaching.

This chapter is positioned within the study of the cultural politics of mathematics education (Valero et al. 2015; Planas and Valero 2016). Bringing together Foucauldian (e.g., Walshaw 2016) and Deleuzian (e.g., de Freitas 2016) analytical strategies, these studies provide an understanding of the cultural and historical constitution of educational practices in mathematics in a multiplicity of interconnected sites, in order to cast light on how mathematics as part of the school curriculum are technologies of power/knowledge, which shape and govern Modern subjectivities and rationalities. In this chapter, this positioning is present in the theoretical landscape adopted, as well as in the analytical strategies deployed. Thus, 'the mathematics teacher' that is referred to here is not to a specific teacher, but rather a notion of mathematics teacher that circulates and is constituted within a discursive network; the mathematics teacher is here considered as a discursive construction fabricated within rationalities and truths that respond to specific spatiotemporal conditions. The chapter deploys a Foucault-inspired discourse analysis (Arribas-Ayllon and Walkerdine 2008; Jørgensen and Phillips 2002), with which a reading of ways of governing the mathematics teacher is opened up, based on what is enunciated as desired and the articulation of a certain form of reasoning and arguing. It is navigating through discursive formations and their resonances to identify forces and different regimes of power/knowledge that determine what is considered true and false regarding the mathematics teacher. But, why is the focus put on the discourses? Through discourses are described rules, divisions and systems of knowledge (Arribas-Ayllon and Walkerdine 2008), in which the notion of the mathematics teacher is constituted and the teacher is drawn. Within discourses, what it means to be a mathematics teacher and what characterizes one, as well as the desired mathematics teacher, are traced.

The empirical materials on which the discourse analysis is deployed consist, on the one hand, of research about the mathematics teacher released within the last five years in scholarly journals (*Journal of Mathematics Teacher Education, Zentralblatt für Didaktik der Mathematik* and *Educational Studies in Mathematics*), and on the other hand, of reports published by the OECD, specifically reports focused on mathematics education and the mathematics teacher, as well as those focused on social welfare and development.

The chapter will follow three movements. Firstly, it discusses the spatiotemporal configuration in which the discourses are shaped, where the mathematics teacher is thought of as a self-regulated professional who always needs to improve. The notion of the '*society of control*' (Deleuze 1992) is used to understand the role of expert knowledge in governance mechanisms. Secondly, it examines the dominant discourses about mathematics teachers and shows the circulation of a promise of salvation based on the need or demand for permanent training. It outlines what characterizes the becoming of the mathematics teacher; a becoming that must undergo continuous change and redefinition with the aim of facing new demands and challenges. And thirdly, it problematizes the constituting and configuration of the outdated mathematics teacher, which embody capitalist and neoliberal rationality. Then, it is established that the success of the mathematics teacher depends on his/her investment in and consumption of permanent training, having a particular effect on the mathematics teacher's ways of acting and being, as well as controlling the teacher through the insatiable search for answering social demands and needs for improvement.

2 Society of Control and Discourses on the Mathematics Teacher

In order to understand how the discursive assemblage of social demands constitutes the mathematics teacher and configures a particular kind of teacher, a reading of the present is opened through the notion of the *society of control* (Deleuze 1992). Firstly, the society of control should be understood not as an overlap, but rather as a displacement of Foucault's *disciplinary societies*. Deleuze "seeks to supplement

Foucault's analysis of disciplinary power by defining new mechanisms of control which, it is suggested, have largely displaced the techniques of power described by Foucault" (Patton 2000, p. 26). Secondly, the society of control should be understood in relation to current capitalist society, in which the focus is not on a particular individual, but on a group of individuals, the mass. The societies characterized by confinement, disciplinary societies, in which the "individual never ceases passing from one closed environment to another, each having its own laws" (Deleuze 1992, p. 3), turn open, in order to enter the market, where the control "is short-term and of rapid rates of turnover, but also continuous and without limit, while discipline was of long duration, infinite and discontinuous" (Deleuze 1992, p. 6). There is an abstraction of all social and personal aspects, which become samples or data, where "the science of the state" or "statistics" (Foucault 1991 [1978]) are put into operation to shape governmentality techniques (Foucault 2010). For example, in international comparative studies, such as reports of the Programme for International Student Assessment (PISA), students involved become samples or data that are studied to formulate discourses as a function of their outcomes, such as, "[o]n average across OECD countries, boys outperform girls in mathematics by eight score points" (OECD 2016a, p. 196). The individual is lost; what is relevant is whether the mass, the particular student group or country involved, achieves what is considered necessary for the input or output of certain categorizations or achievement and development levels. The OECD (2016a) asserted that

> [o]n average across OECD countries, only 2.3% of students attain Level 6 [score higher than 669 points in PISA]. More than one in ten students perform at this level in Singapore (13.1%) and Chinese Taipei (10.1%). In B-S-J-G (China), Hong Kong (China), Japan, Korea and Switzerland, between 5% and 10% of students attain proficiency Level 6. In 30 participating countries and economies, between 1% and 5% of students perform at this level, in 21 countries/economies, between 0.1% and 1% of students performs at Level 6, and in 12 other countries/economies, fewer than one in one thousand students (0.1%) performs at Level 6. (pp. 193–194)

On the basis of students' performance on standardized tests, diverse countries formulate new requirements for teachers and schools, enunciating what to do and how to do it, conducting the conduct of mass:

> [M]ore and more countries are looking beyond their own borders for evidence of the most successful and efficient education policies and practices. [...] PISA allows governments and educators to identify effective policies that they can then adapt to their local contexts. (OECD 2016b, p. 3)

Currently, the teaching and learning of mathematics is a cornerstone of modernity, social progress and development, since mathematics provides the language of science and technology, as well as a rationality desired for subjects. It is possible to see that mathematical development is the foundation of much of the scientific and technological activity that distinguishes advanced from those less advanced. The value of mathematics is "a result of the formal place mathematics occupies within late capitalism" (Pais 2013, p. 20). In this fashion, it has been enunciated that school-level mathematics is relevant, since it "can enhance "personal and social capability" by providing opportunities for initiative taking, decision making, communicating processes and findings" (OECD 2015, p. 99). The idea circulates that the proper acquisition of mathematics skills, especially numeracy, is needed for citizens to achieve their full potential and development, enabling them to excel and have better lives. Moreover, it is recognized that "mathematics education inserts children in the great Modern narrative of knowledge for problem solving [..., causing] children to see themselves as agents who can bring about change in the world and so contribute to the betterment and progress of society" (Valero and Knijnik 2015, p. 35). But, it is not enough for a person merely to have mathematical knowledge. Rather, he/she has to have the skill to put such knowledge into operation in diverse contexts, what is called *mathematical literacy* (see OECD 2014a).

Research focused on mathematics teachers has been developed from diverse frameworks and approaches. This kind of research has gained relevance in mathematics education:

> As the field of mathematics education grows so too do the research methods used to study the field. In the special area of teacher education, the last decade has witnessed a substantial increase in attention. New perspectives and new methodologies have been constituted and new research techniques established. (Gellert et al. 2013, p. 327)

Literature reviewed by Goldsmith et al. (2013) shows that several lines of research have been developed in the professional learning of practicing teachers of mathematics. They identify several crosscutting themes in the literature, proposing that growth in one aspect of teachers' knowledge and practice may promote growth in other areas. The studies analyzed were clustered into nine categories or areas: teachers' identity, beliefs, and dispositions; teachers' instructional practice; mathematical content of lessons; changes in classroom discourse; promoting students' intellectual autonomy; teachers' collaboration/community; teachers' attention to student thinking; mathematics content knowledge; and curriculum and instructional tasks.

The research has become a mechanism of control. Through research, it is possible to know whether teachers have the quality levels required in their training and practices to ensure certain outcomes. Within research, the teacher's ways of acting and being are directed on the basis of what is enunciated as the desired mathematics teacher and what characterizes this ideal teacher. Through this mechanism, the mathematics teacher is framed in the logic of competition and comparison. Teachers have to compete and be compared against each other, in order to show that they are competent and effective, that they are better than others, and that they have everything necessary for reducing the gap between themselves and the desired teacher. Competence and comparison are configured as important elements of the neoliberal agenda, which are promoted as a means for improvement. Competence and comparison are set as a way of governing the conduct of the mathematics teacher, as well as a way of life and of thinking, constituting truths and discourses regarding what is possible and desired. It is believed that competence and comparison contribute positively to increasing teachers' competitiveness and effectiveness, since through these they can know their strengths and weaknesses, as well as what they need to improve to become successful and to be regarded as quality teachers.

Nowadays, it is impossible to think about mathematics teachers without considering their social connections or implications. Mathematics teachers and their education have "links and connections to many other fields within and outside mathematics education" (da Ponte 2013, p. 489). Therefore, it has been recognized, within different levels of the social sphere and by different agents, that the teachers play "a unique role as experts who provide opportunities for students to engage in the practices of the mathematics community" (Bleiler et al. 2013, p. 105). Also, navigating through the discourses that circulate about the mathematics teacher, it is possible to see how this teacher is constituted in relation to social interests and demands. For example, a general demand drawn from different levels of social spheres is to improve people's quality of life and well-being. This demand has put into operation dispositives and forces that seek to improve educational achievement, since it is believed that high educational achievement helps to improve the quality of life and the well-being of a person (OECD 2016a, b). Along with this, there is also the demand for an effective and competitive teacher, a highly qualified individual who possesses excellent teacher training and, thus, up-to-date professional knowledge and skills, to address challenges that arise in the search for ways to improve educational achievement. So, the mathematics teacher is constituted as a product of discursive assemblage of social demands and interests, in which the idea is set and circulated that better mathematical achievement will lead to the improvement of living conditions at individual and national levels.

Inside the circulating discourses are deployed a large number of arguments about the things that the teacher must improve to become a "good" teacher, in other words, an effective, competitive and successful teacher. The idea of a "good" teacher promotes, on the one hand, the setting of truths, rationalities, discourses and subjectivities; and on the other hand, defines the space for what is allowed and prohibited, by configuring a network of forces and enunciations to which the mathematics teacher is subjected. However, the circulating discourses are constantly reformulating what is considered a "good" teacher. Over recent decades, there have been great efforts to improve teaching and develop teachers (Huang and Shimizu 2016). Within discourses circulate a long list of qualities and capacities that the teacher needs to develop in order to fulfill different aspects that lead him/her to be recognized as an effective, competitive and successful teacher. There is an emphasis on communication, detecting students' learning, knowledge, tools and skills that the teacher needs, (among others). For example:

Teachers need to be able to notice children's means of communicating their reasoning in order to respond appropriately to enhance children's reasoning and communication of their mathematical thinking. (Bragg et al. 2016, p. 524)

Teachers need specific knowledge and affect-motivational skills to diagnose students' learning during class. (Hoth et al. 2016, p. 44)

The kind of education needed today requires teachers to be high-level knowledge workers who constantly advance their own professional knowledge as well as that of their profession. Teachers need to be agents of innovation not least because innovation is critically important for generating new sources of growth through improved efficiency and productivity. (OECD 2012, p. 36)

Teachers need to understand (1) the conceptual principles and the development of the ideas underlying a concept; (2) strategies, representations and misconceptions; (3) meaningful distinctions, definitions and multiple models; (4) coherent structure—recognizing that there is a pattern in the development of mathematical ideas as a concept becomes more complex; and (5) bridging standards—understanding that there might be gaps between standards and knowing what underlying concepts are in between to bridge the gaps between the standards. (Suh and Seshaiyer 2014, p. 209)

The enunciations of what the mathematics teacher needs or has to improve, continually expresses the double gestures (Popkewitz 2008b) of hope and fear. Society hopes that the mathematics teacher will become a competitive and effective teacher who embodies ways of thinking and acting, promoting a particular rationality. But, at the same time, it is enunciating the feared mathematics teacher, which society does not desire, a teacher that will hinder society in what it seeks to achieve. Within circulating discourses have constituted that the expert knowledge shapes a promise of salvation based on permanent training, in order that mathematics teachers do not become an undesired and valueless teacher—an outdated teacher. Moreover, within the expert knowledge, a notion of the teacher as a desired, self-regulated and productive teacher is constituted. Then, those teachers, whose subjectivities are adjusted with what is desired, will be saved, while those who do not achieve the desired adjustment, will be outside the system and will be excluded. Within double gestures, the becoming of the mathematics teacher is drawn, revealing the features of the successful mathematics teacher. The becoming of the mathematics teacher is characterized by constant changes, which are a function of the emergence of new social demands and requirements. Nowadays, nothing is finished; in other words, everything is in a state of permanent becoming, and the mathematics teacher is no exception. In the words of Deleuze (1992),

[i]n the disciplinary societies one was always starting again (from school to the barracks, from the barracks to the factory), while in the societies of control one is never finished with anything—the corporation, the educational system, the armed services being metastable states coexisting in one and the same modulation, like a universal system of deformation. (p. 5)

3   A Promise of Salvation

The circulating demand of better mathematics teachers, effective and competitive teachers who can overcome new social challenges and requirements and who are always adapting and improving, constitutes a network of discourses in which the becoming of mathematics teachers is shaped. The becoming of teachers is in perpetual change and their quality has been constantly questioned. From different perspectives, theories and methodologies have enunciated that "teachers are key to increasing educational quality" (Luschei and Chudgar 2015, p. 3), as well as that mathematics teachers have diverse deficits, for example, in their knowledge, resources for teaching or outcomes, among others. Francis (2015) shows that a low sense of efficacy and perceived lack of preparation are two factors that impact how teachers teach and interact with their students. Hence, "teachers in classrooms are the main factor in bringing about improvement in students' outcomes" (Callingham et al. 2015, p. 552). Within discursive formations circulate as truth the notion that the mathematics teacher needs to be in a constant state of improvement with regard to all his/her professional and personal aspects in order to become a successful, effective and competitive teacher. In this vein, it is enunciated that

> teachers should gradually improve their practice over time by engaging in systematic analysis of the effects of instruction on student learning. (Spitzer et al. 2011, p. 68)

> [It] highlights the importance of teachers and school and system leaders increasingly taking responsibility for improving the enactment of the sequence, and for drawing on the underlying principles in various aspects of their practice. (Cobb and Jackson 2015, p. 1029)

> The present study showed that French teacher training in the concepts of attribute and measurement is insufficient and therefore must be expanded. First, it is necessary to improve future teachers' mastery of these concepts, which means developing their attribute/measurement SCK [specialised content knowledge]. (Passelaigue and Munier 2015, pp. 333–334)

> Future studies are needed to explore strategies to help teachers improve their pedagogical design capability and flexibility to handle emerging events in the classroom. (Cai et al. 2014, p. 279)

These statements articulate discourses about not only what teachers should improve or change but also about whom they should become, by illustrating the desired teacher that society needs or demands. Moreover, it is recognized that initial teacher education is insufficient to satisfy the new challenges and changes that society sets. There is the fear of failure, as well, that the mathematics teacher becomes outdated. Current social changes happen fast, promoting new demands and requirements that the mathematics teacher has to face. In this context, various aspects of the mathematics teacher run the risk of becoming inefficient, lacking in usefulness or value because they respond to a reality that has already changed. Societal demands draw a network of forces, which delineate and shape what the mathematics teacher must be and do. But, these demands are in movement, constantly (re)forming from new interests, events, needs and opportunities. Consequently, the mathematics teacher is continually redefined, including his/her practices, training and performances. Mathematics teachers must quickly adapt, changing and improving all aspects of themselves that are identified as necessary to be effective and competitive. The outdated aspect(s) of these teachers are constructed as a real problem for society, not just because such teachers could have difficulty in finding or keeping a job, but also because they could hamper the achievement of educational objectives or desired standards.

Studies and discourses about the mathematics teacher seek to put into operation arguments based on the empirical evidence produced by research, to promote the best way of redesigning and re-planning the teachers, their (re)training, practices, and work, delineating what it means to be a mathematics teacher, and what is considered urgent. They contend that "the investigation experience is very intense and has a high transformation and learning potential for the participants" (da Ponte et al. 2017, p. 292). This experience creates opportunities for developing new knowledge and skills, as well as promoting changes in the participants and in their identity, values, beliefs and ways of being, among other respects (op. cit.). Within circulating discourses, practices are constituted, as are repertoires of techniques and knowledge that are considered desirable and those that respond from evidence in the best possible way to social requirements, or that have been shown to be successful. The fabrication of the effective and competitive mathematics teacher, as well as his/her training, has been part of the agenda in recent years of mathematics education research and studies by international agencies, in which it is possible to see how "policy makers put pressure on teachers to perform according to their own pedagogical and curricular demands" (Lerman 2012, p. 188). Cochran-Smith and Villegas (2015) state that current research on the preparation of prospective teachers is based on two broad questions: the policy question, which involves issues of effectiveness as well as accountability; and the learning question, which involves the issue of how (prospective) teachers learn to teach in the 21st century. Regarding the quality of teachers, it is possible to see efforts at various levels, such as the reformulation of mathematics teacher training (see Lerman 2012) or the configuration of international policies to attract the best students to become teachers, retain the best teachers, increase teacher salaries, enhance working conditions for teachers and reward teachers that join schools with the greatest needs, with the aim of securing better-quality teachers (Luschei and Chudgar 2015; OECD 2005, 2012, 2014b). It is acknowledging the central role of teachers for the social and personal development, as well as for a better future.

The permanent training embodies the hope of an effective and competitive teacher, as well as the fears of those who are not seeking to improve their practices or not achieving the desired levels. The narratives of permanent training shape a promise of salvation regarding the failure or incompetence of mathematics teachers. The search for constant improvement entails enunciating the desired and undesired teacher, labeling one as good or bad teacher according to his/her outcomes, what they know to do, and their updated repertoire of techniques and knowledge. Permanent training is configured as a way of facing what are recognized as deficient aspects of the teacher, as well as a mechanism of continuous control of his/her academic and professional development, in order to ensure the quality of the teacher and encourage his/her standardization. Currently, it is impossible to think of the becoming or professional development of the mathematics teacher outside the process of permanent training. The permanent training is the action by which the mathematics teacher becomes "other" while continuing to strive to be "what is", governing the mathematics teacher and his/her desires, fears, attitudes and ways of being, in order to produce the desired teacher:

The idea of permanent training is a way of maintaining control of a never-ending process for the teacher. The idea of permanent training is operating as part of a dispositive by setting diverse forms of control, discourses, and forces. Consequently, the mathematics teacher is condemned to be incomplete and to have constant deficits to overcome, since society and the market will always be setting new requirements, demands, and urgencies that the teacher must face. (Montecino and Valero 2017, p. 150)

Permanent training becomes the cornerstone for the realization of current effective and competitive teachers, shaping teachers' conduct by inculcating into them new ways of understanding their professional development, work, desires, and ways of acting and being. Socially, it seeks to fabricate a trained teacher who is able to improve through self-regulation and the pursuit of their interests. As Callingham et al. (2015) put it,

the teachers […] had made a collective decision to focus on and improve their personal numeracy (or quantitative reasoning) in ways that would allow them to challenge their students appropriately. (p. 558)

Consequently, permanent training implies a promise to reshape teachers in such a way that they can address new social demands and changes. The narratives about the competitive and effective mathematics teacher have subjected the teacher to a process of permanent training, in which the shaping of teachers' subjectivities takes place. Permanent training ensures continuous control of academic and professional development and practices of the mathematics teacher. Also, the permanent training, which is a response to urgent needs, is required for providing "quality control" in teaching, and for favoring the standardization of teachers. Within these discourses there seems to be a need for the existence of teachers who fail or do not satisfy what is desired, and in this way, it is possible to keep open and active, a market for permanent training.

4 The Outdated Mathematics Teacher

The search for ways to improve the mathematics teacher has become the central aim of studies that have as their focus the mathematics teacher and his/her training, practices, knowledge, outcomes and other related subjects. According to circulating discourses, it seems that the teacher cannot achieve the quality levels that are desired, which constantly change. Expert knowledge sets and contributes to new regimes of truth about teachers, circulating the idea that all mathematics teachers have opportunities and possibilities for improving through permanent training. There is the constant risk that mathematics teachers become outdated if they are not updated and adjusted in a timely manner.

As a result of constant social changes, demands and interests, the competitiveness and effectiveness of the mathematics teacher are destined to have a limited useful life. For example, the repertoire of techniques that a teacher should currently have, or is currently demanded to have, will be different in the future. This change is clear with the introduction of new technologies—e.g. calculators, computers, motion sensors—in the classroom since the 1980s. Teachers had to acquire new knowledge and techniques, in order to abandon the exclusive use of the blackboard and to integrate the new technologies into their practices.

Plenty of research has argued about what characterizes a good teacher and what teachers need to improve their practices, knowledge and achievements. For example, Ertle et al. (2016) enunciated that

> teachers must ultimately learn to develop and conduct their own formative assessments. To do this, teachers need to understand their students' development of mathematical knowledge and thinking as well as the techniques (e.g., observation and interviewing) for supporting and assessing children's knowledge in a formative way. (p. 977)

Apparently, all teachers have the same opportunities, access and possibilities for improving. But, access to the permanent training is not for all. In the last years, there has been a turn from demands for obedience on the part of teachers towards demands that teachers be adaptable, flexible, versatile and entrepreneurial.

> So teaching staff nowadays also need the competences to constantly innovate and adapt; this includes having critical, evidence-based attitudes, enabling them to respond to students' outcomes, new evidence from inside and outside the classroom, and professional dialogue, in order to adapt their own practices. (European Comission 2013, p. 7)

The constant enunciation of the new repertoire of techniques and knowledge that the mathematics teacher must have or develop to be considered a successful teacher becomes a form of governing, constituting notions about the mathematics teacher that organize their practices. These notions produce truths and discourses that determine what it is possible to say, do and be, and a form of governing that conducts the mathematics teacher into consumerism and accumulation. The teacher must invest in his/her professional development and consume new training offers to achieve the desired level of effectiveness and competitiveness, accumulating more and more teaching methods, knowledge, achievements, retraining and so on. It establishes that consumerism and accumulation of professional development are the only ways for improving and not becoming outdated, as well as a means for self-regulating and circulating in a society that functions according to the principles of the market. The role of the mathematics teacher seems to be reduced to knowing what to consume to satisfy social demands and needs, while at the same time, it seems that the teacher has developed the desire to possess newer and better techniques and knowledge, in order to be or to stay effective and competitive, since his/her training and preparation are not enough. "Yet despite widespread recognition that teachers need to learn more in order for students to learn more, there is little consensus about what it is that teachers should be learning" (Lewis et al. 2015, p. 448). Constant changes make it more complex to define or establish what it is that teachers should learn, have or develop. The only certainty is that teachers have to improve since they have or will be in deficit.

Rapid social changes are constantly increasing the gap between the actual mathematics teacher and the desired teacher. The mathematics teacher has to work at reducing this gap, by managing his/her training and updates, in other words, the teacher needs to become a sort of enterprise for him/herself (Popkewitz 2008b), in order to pursue his/her better professional and academic development. Hence, in the words of Popkewitz (2008a), within a neoliberal society, the subject becomes an "individual who is continually pursuing knowledge and innovation in a never ending chase for the future" (op. cit. p. 310), which applies to the fabrication of the mathematics teacher nowadays. Teachers should perceive themselves as 'agents' (Foucault 2009) who are responsible for their own improvement and professional development. Teachers have to manage their professional development and investments, in order to improve and ensure their quality—teachers are responsible for themselves—and to keep up in the system. "[Q]uality is conflated with measurable progress within neoliberalism where national progress (economic growth and competitiveness) is matched with individual progress (personal growth and self-fulfilment)" (Llewellyn and Mendick 2011, p. 56). Circulating the idea that the more mathematics teachers are of quality, the greater the social progress. The individual ambition of the mathematics teacher (the searching for being the best or improving his/her quality) becomes a key element for neoliberalism.

The mathematics teacher is subjected to a market logic based on "individual capitalization" (Foucault 2010), indebtedness and economic investment, that seeks the normalization, standardization and control of all aspects of academic and professional development of the mathematics teacher. Standardization becomes a mechanism for segregation and differentiation since the standards are used as a way to measure the effectiveness and efficiency of the teacher. The mathematics teacher is configured through political, economic and social interests, which are focused on competitiveness, effectiveness and qualifications of the teacher with the aim of satisfying new challenges that the market sets. The market shapes the conditions for the mathematics teacher to become part of the idea of globalization, social progress and the competitive logic of current societies. The market determines what is valid, permitted and desired for the mathematics teacher. In other words, the market determines what must be the academic and professional development of the mathematics teacher, along with his/her training, practices, achievements and knowledge.

For neoliberalism and the society of control, the professional freedom of the mathematics teacher is a function of the market, and all areas of the mathematics teacher are cast in economic terms, namely in terms of costs and benefits. Mathematics teachers are led to act under their own individual interests, having the freedom of choosing what they want to consume with the aim of improving. Hence, the success of the mathematics teacher is a function of his/her capacity to invest and consume wisely. Moreover, the configuration and reshaping of the mathematics teacher are based on market principles; supply and demand promotes the mathematics teacher's ways of being and acting, as well as their constant fear of becoming outdated, losing value and effectiveness.

The permanent training not only promotes continuous updating of mathematics teachers, but also inserts them into continuous practices of comparison and competition. These practices embody capitalist and consumerist ideas, whereby the mathematics teacher is denaturalized of his/her particularities, framing the becoming of the mathematics teacher as an outdated teacher who is always in debt and lacking something. Teachers are governed and govern themselves in order to optimally satisfy the social demands conducting their ways of thinking and acting. Furthermore, because the mathematics teacher governs others, he/she plays a part in the furtherance of a capitalist rationality and of practices that nowadays are believed essential and natural for achieving personal and social success.

> Governing people is not a way to force people to do what the governor wants; it is always a versatile equilibrium, with complementarity and conflicts between techniques which assure coercion and processes through which the self is constructed or modified by himself. (Foucault 1993, p. 204)

The expert knowledge puts into operation and (re)produces arguments for setting permanent training as the main means for improving and ensuring the quality of the teacher, despite all the evidence which shows that failure is unavoidable. Within research, there is a production of new forms of (re)training and methods for improving the mathematics teacher, each carrying the promise of being the Holy Grail for constituting the successful teacher, promoting in the teacher the desire for consuming it, in order to improve. Currently, mathematics teachers must engage in a constant training and improvement process. They should acquire not only the skills and characteristics of the desired teacher, but also certain ways of acting. In other words, teachers are subjected to practices that shape their conduct to achieve the qualifications needed and to improve, in order to be considered effective and competitive.

5   Conclusion

The pursuit of improving of each aspect of the teacher has, in large part, become the focus of studies regarding the mathematics teacher, formulating discourses around better practices, knowledge and repertoires of techniques that the teacher must have to achieve successful outcomes, to be considered competent and effective, and to not become outdated—discourses in which truths regarding what characterizes the desired mathematics teacher are (re)produced. The enunciation of new practices, knowledge and repertoires that the mathematics teacher must have or improve upon, become part of the technologies for governing the mathematics teacher. In the context of current societies of control, as Deleuze has pointed out, the rendering of subjects to the myriad of control mechanisms effectively turns people into the perfect subjects/objects of expansive, late capitalist markets. In other words, current technologies of conducting the conduct of individuals and societies bring us all, sometimes subtly and sometimes forcefully, into a world of value, consumption and marketization. Notions of the mathematics teacher—and connected subjectivities—are no exception.

Who would want to be educated by a bad or outdated teacher? Nobody. The desire for good mathematics teachers makes it almost impossible to problematize and question the intrinsic goodness of the "necessity" for improving teachers. The "necessity" of permanent training feeds the illusion that there is a way for the mathematics teacher to become the desired teacher or at least to come close to it; that the teacher is not alone; that there are people working to help him/her; and that all the teacher's deficits or problems can be fixed through research, constituting new regimes of the production of truth, in which truths are (re)produced from a scientific discourse. But permanent training does not only promote continuous updating of mathematics teachers, it also pushes them into continuous practices of comparison and competition. These practices frame the teacher's possibilities within a capitalist and consumerist logic, since that comparison and competition leads the teacher to think that pursuing his/her personal interest, investing in him/herself through permanent training or any other aspect that can be capitalized will indeed result in added value and will lead to becoming the best possible modified version of a teacher. In the discourse of necessity in the market, the mathematics teacher surrenders to permanent training because he/she desires certain effects, seeking a personal and social benefit.

Nevertheless, the very same logic of societies of control operate that the desire for success often turns into failure. Despite the permanent training, the mathematics teacher is condemned to fail in acquiring new qualifications, and as a result he/she will be doomed to become outdated. The fabrication of the mathematics teacher seems to be framed in a narrative of *planned obsolescence,* since, on the one hand, the mathematics teacher is purposefully destined to become outdated, and on the other hand, the teacher is shaped to always desire the consumption of something new. Nowadays, most technologies and devices are designed with a short useful life, and the mathematics teacher is no exception. Planned obsolescence is the outcome of the decision that a product should no longer be functional or desirable after a predetermined period, determining its intrinsic durability (Cooper 2011). In other words, it is a purposeful strategy of built-in product design that reduces the product's useful life, triggering in consumers the desire for a newer and better product of the same type. Planned obsolescence is one of the great inventions of recent electronic and digital technologies that operate with a high speed of change in products, which consumers perceive as necessary and highly desirable. The mathematics teacher is fabricated within a production strategy, the planned obsolescence, which seeks to stimulate economic activity (Cooper 2011).

In this way, the mathematics teacher becomes a part of a market logic, whereby issues such as professional qualification, effectiveness and levels of success (among others) take on particular relevance since they become control mechanisms, which can be quantifiable, for the identification of those who fulfill the quality requirements or standards. The mathematics teacher becomes a subject who never stops desiring more qualifications and training, since these are perceived as never being enough. The constant interplay between the discourse of salvation and the becoming of the outdated mathematics teacher has an effect of power on his/her ways of acting and being. The mathematics teacher is controlled by the insatiable search for responses to social demands, as well as the search for improvement. And this in itself becomes the very same condition for his/her being professional. It seems as if the vicious circle between the desire for more qualifications and the fact that qualifications become obsolete creates a kind of illusion about what may be possible for teachers to in fact achieve. Learning to be a teacher whose competence, qualifications and professionalism are always insufficient raises the ethical question of which kinds of teachers are being formed. Is "a teacher in debt" (Montecino and Valero 2017, p. 150) an acceptable and desirable subject?

But at this point, one might ask, what happens with resistance? Are there no other possibilities to envision and conceive of mathematics teachers? According to Foucault (1980), where there is power, there is always resistance, and resistance is never exterior to power. Moreover, discourse transmits and produces power, but it also undermines and exposes it (op. cit.). The mathematics teacher freely participates and navigates within mechanisms of power. However, the resistance cannot be total because when the mathematics teachers do not adapt, they become outdated. Thus, within the current configuration of a capitalist model, the mathematics teacher is shaped as product and agent, as well as a professional who must be effective and competitive. The mathematics teacher, as a professional, is required to know how to stay up to date. The resistance is made difficult, since if mathematics teachers do not improve and if they do not consume pursuing their interests, they will become valueless and outdated subjects. Moreover, the capitalist model resignifies all that is considered new, different and feared, including the resistance, in its own terms. Therefore, mathematics teachers are subjected to the flow of their society. Although possible, the cracks that may open opportunities for resistance, seem at the moment, quite difficult to see and from which to make profit.

References

Arribas-Ayllon, M., & Walkerdine, V. (2008). Foucauldian discourse analysis. In C. Willig & W. Stainton-Rogers (Eds.), *The SAGE handbook of qualitative research in psychology* (pp. 91–108). London: SAGE Publications.

Bleiler, S. K., Thompson, D. R., & Krajčevski, M. (2013). Providing written feedback on students' mathematical arguments: Proof validations of prospective secondary mathematics teachers. *Journal of Mathematics Teacher Education, 17*(2), 105–127.

Bragg, L. A., Herbert, S., Loong, E. Y.-K., Vale, C., & Widjaja, W. (2016). Primary teachers notice the impact of language on children's mathematical reasoning. *Mathematics Education Research Journal, 28*(4), 523–544.

Cai, J., Ding, M., & Wang, T. (2014). How do exemplary Chinese and U.S. mathematics teachers view instructional coherence? *Educational Studies in Mathematics, 85*(2), 265–280.

Callingham, R., Beswick, K., & Ferme, E. (2015). An initial exploration of teachers' numeracy in the context of professional capital. *ZDM Mathematics Education, 47*(4), 549–560.

Cobb, P., & Jackson, K. (2015). Supporting teachers' use of research-based instructional sequences. *ZDM Mathematics Education, 47*(6), 1027–1038.

Cochran-Smith, M., & Villegas, A. M. (2015). Studying teacher preparation: The questions that drive research. *European Educational Research Journal, 14*(5), 379–394.

Cooper, T. (2011). Planned obsolescence. In D. Southerton (Ed.), *Encyclopedia of consumer culture* (p. 1096). Thousand Oaks, California: SAGE Publications Inc.

da Ponte, J. P. (2013). Theoretical frameworks in researching mathematics teacher knowledge, practice, and development. *Journal of Mathematics Teacher Education, 16*(5), 319–322.

da Ponte, J. P., Santos, L., Oliveira, H., & Henriques, A. (2017). Research on teaching practice in a Portuguese initial secondary mathematics teacher education program. *ZDM Mathematics Education, 49*(2), 291–303.

de Freitas, E. (2016). Gilles Deleuze. *Alternative theoretical frameworks for mathematics education research: Theory meets data* (pp. 93–120). Cham: Springer International Publishing.

Deleuze, G. (1992). Postscript on the societies of control. *October, 59*, 3–7.

Ertle, B., Rosenfeld, D., Presser, A. L., & Goldstein, M. (2016). Preparing preschool teachers to use and benefit from formative assessment: The birthday party assessment professional development system. *ZDM Mathematics Education, 48*(7), 977–989.

European Commission. (2013). *Supporting teacher competence development for better learning outcomes*. European Commission.

Foucault, M. (1980). *Power/knowledge: Selected interviews and other writings 1972–1977*. New York: Pantheon Books.

Foucault, M. (1991 [1978]). Governmentality. In G. Burchell, C. Gordon, & P. Miller (Eds.), *The Foucault effect: Studies in governmentality* (pp. 87–104). Chicago, IL: University of Chicago Press.

Foucault, M. (1993). About the beginning of the hermeneutics of the self: Two lectures at Dartmouth. *Political Theory, 21*(2), 198–227.

Foucault, M. (2009). *Security, territory, population: Lectures at the Collège de France 1977–1978* (Vol. 4). In M. Senellart, F. Ewald, & A. Fontana (Eds.). New York: Macmillan.

Foucault, M. (2010). *The birth of biopolitics: Lectures at the Collège de France, 1978–1979*. In G. Burchell (Ed.). New York: Picador.

Francis, D. I. C. (2015). Dispelling the notion of inconsistencies in teachers' mathematics beliefs and practices: A 3-year case study. *Journal of Mathematics Teacher Education, 18*(2), 173–201.

Gellert, U., Hernández, R., & Chapman, O. (2013). Research methods in mathematics teacher education. In M. A. Clements, A. J. Bishop, C. Keitel, J. Kilpatrick, & F. K. S. Leung (Eds.), *Third international handbook of mathematics education* (Vol. 27, pp. 327–360). New York: Springer.

Goldsmith, L. T., Doerr, H. M., & Lewis, C. C. (2013). Mathematics teachers' learning: A conceptual framework and synthesis of research. *Journal of Mathematics Teacher Education, 17*(1), 5–36.

Gutierrez, R. (2013). The sociopolitical turn in mathematics education. *Journal for Research in Mathematics Education, 44*(1), 37–68.

Hoth, J., Döhrmann, M., Kaiser, G., Busse, A., König, J., & Blömeke, S. (2016). Diagnostic competence of primary school mathematics teachers during classroom situations. *ZDM Mathematics Education, 48*(1), 41–53.

Huang, R., & Shimizu, Y. (2016). Improving teaching, developing teachers and teacher educators, and linking theory and practice through lesson study in mathematics: An international perspective. *ZDM Mathematics Education, 48*(4), 393–409.

Jacob, B., Frenzel, A. C., & Stephens, E. J. (2017). Good teaching feels good—But what is "good teaching"? Exploring teachers' definitions of teaching success in mathematics. *ZDM Mathematics Education, 49*(3), 461–473.

Jørgensen, M., & Phillips, L. J. (2002). *Discourse analysis as theory and method*. London: SAGE Publications.

Lerman, S. (2012). Mapping the effects of policy on mathematics teacher education. *Educational Studies in Mathematics, 87*(2), 187–201.

Lewis, J., Fischman, D., & Riggs, M. (2015). Defining, developing, and measuring "Proclivities for Teaching Mathematics". *Journal of Mathematics Teacher Education, 18*(5), 447–465.

Llewellyn, A., & Mendick, H. (2011). Does every child count? Quality, equity and mathematics with/in neoliberalism. In B. Atweh, M. Graven, W. Secada, & P. Valero (Eds.), *Mapping equity and quality in mathematics education* (pp. 49–62). Dordrecht: Springer Netherlands.

Luschei, T., & Chudgar, A. (2015). *Evolution of policies on teacher deployment to disadvantaged areas*. Education for all, Global Monitoring Report.

Montecino, A. (2017). The fabrication of the mathematics teacher as neoliberal subject. *Doctoral dissertation*. Aalborg University, Denmark (Manuscript in preparation).

Montecino, A., & Valero, P. (2017). Mathematics teachers as products and agents: To be and not to be. That's the point! In H. Straehler-Pohl, N. Bohlmann, & A. Pais (Eds.), *The disorder of mathematics education: Challenging the sociopolitical dimensions of research* (pp. 135–152). Cham: Springer International Publishing.

OECD. (2005). *Teachers matter: Attracting, developing and retaining effective teachers*. Paris: OECD Publishing.

OECD. (2012). *Preparing teachers and developing school leaders for the 21st century: Lessons from around the world*. In A. Schleicher (Ed.). Paris: OECD Publishing.

OECD. (2014a). *Education at a glance 2014: OECD indicators*. Paris: OECD Publishing.

OECD. (2014b). How much are teachers paid and how much does it matter? *Education Indicators in Focus, 21*.

OECD. (2015). *Skills for social progress: The power of social and emotional skills*. OECD skills studies. Paris: OECD Publishing.

OECD. (2016a). *PISA 2015 assessment and analytical framework: Science, reading, mathematic and financial literacy*. Paris: PISA, OECD Publishing.

OECD. (2016b). *PISA 2015 results (volume I): Excellence and equity in education*. Paris: PISA, OECD Publishing.

Pais, A. (2013). An ideology critique of the use-value of mathematics. *Educational Studies in Mathematics, 84*(1), 15–34.

Passelaigue, D., & Munier, V. (2015). Schoolteacher trainees' difficulties about the concepts of attribute and measurement. *Educational Studies in Mathematics, 89*(3), 307–336.

Patton, P. (2000). *Deleuze and the political*. London: Routledge.

Planas, N., & Valero, P. (2016). Tracing the socio-cultural-political axis in understanding mathematics education. In Á. Gutiérrez, G. C. Leder, & P. Boero (Eds.), *The second handbook of research on the psychology of mathematics education: The journey continues* (pp. 447–479). Rotterdam: Sense Publishers.

Popkewitz, T. (2008a). Education sciences, schooling, and abjection: Recognizing difference and the making of inequality? *South African Journal of Education, 28*(3), 301–319.

Popkewitz, T. S. (2008b). *Cosmopolitanism and the age of school reform: Science, education, and making society by making the child*. New York: Routledge.

Ribeiro, C. (2011). "Thought of the outside", knowledge and thought in education: Conversations with Michel Foucault. *Educação e Pesquisa, 37*(3), 613–628.

Skilling, K., Bobis, J., Martin, A. J., Anderson, J., & Way, J. (2016). What secondary teachers think and do about student engagement in mathematics. *Mathematics Education Research Journal, 28*(4), 545–566.

Spitzer, S. M., Phelps, C. M., Beyers, J. E. R., Johnson, D. Y., & Sieminski, E. M. (2011). Developing prospective elementary teachers' abilities to identify evidence of student mathematical achievement. *Journal of Mathematics Teacher Education, 14*(1), 67–87.

Suh, J., & Seshaiyer, P. (2014). Examining teachers' understanding of the mathematical learning progression through vertical articulation during Lesson Study. *Journal of Mathematics Teacher Education, 18*(3), 207–229.

UNESCO. (2007). *Education for all by 2015-will we make it?* Paris: UNESCO Publishing and Oxford University Press.

Valero, P., Andrade-Molina, M., & Montecino, A. (2015). Lo político en la educación matemática: De la educación matemática crítica a la política cultural de la educación matemática. *Revista Latinoamericana de Investigación en Matemática Educativa, 18*(3), 287–300.

Valero, P., & Knijnik, G. (2015). Governing the modern, neoliberal child through ICT research in mathematics education. *For the Learning of Mathematics, 35*(2), 34–39.

Walshaw, M. (2016). Michel Foucault. *Alternative theoretical frameworks for mathematics education research: Theory meets data* (pp. 39–64). Cham: Springer International Publishing.

Part IV
Policy and the Sociopolitical in Mathematics Education

Mathematics Curricula: Issues of Access and Quality

Tamsin Meaney

Abstract In the last forty years, it seems that discussions about inequality, power, access and identity have simultaneously become more prominent in mathematics education curricula, whilst also being subordinated to wider neo-liberal discourses of competition and accountability. In this paper, issues to do with access and quality are linked to the mechanism that determines what constitutes mathematics education for specific groups of children. Using Bernstein's pedagogical device, it is possible to see how the control of official knowledge affects who has access to what kind of mathematics learning opportunities. At the same time, contestation about what should be official knowledge also allows alternative possibilities to be raised. The challenge for those interested in providing higher quality mathematics education to all groups of students is how to make use of the possibility for "unthinkable" knowledge.

Keywords Curriculum · Equity · Neo-liberalism · Bernstein · Pedagogic device

1 Introduction

In recent times, educational policy-making in nation states, including the production of curricula,[1] is considered to be submerged into global trends (Atweh and Clarkson 2002a). Yet, perspectives of how mathematics should be taught and learnt, a regular feature of many curricula, have a long history of being affected by global discourses, that are then recontextualised into local settings (see for example, Meaney 2014; Wake and Burkhardt 2013). Following my previous work on the

---

[1]In this chapter, I use the definition of curriculum from Kamens and Benavot (1991), "Throughout this article the term "curriculum" refers to the official subject matter that has been transmitted by national educational administrators or system-level authorities to be taught in local schools" (p. 174).

T. Meaney (✉)
Western Norway University of Applied Science, Bergen, Norway
e-mail: Tamsin.Jillian.Meaney@hvl.no

© Springer International Publishing AG 2018
M. Jurdak and R. Vithal (eds.), *Sociopolitical Dimensions of Mathematics Education*, ICME-13 Monographs, https://doi.org/10.1007/978-3-319-72610-6_10

impact of global discourses on local educational practices, I explore how two recent discourses, one to do with equity in education and the other, a neo-liberal discourse, about the need for accountability through competition, interact in mathematics curricula decision making. These two discourses are investigated because of their potential impact on student and educator perceptions on how inequality, power, access and identify are integrated into mathematics education. A suggestion for the mechanism leading to convergence of curricula is put forward.

A global discourse can contribute both to describing as well as constituting a recognisable social event (Fairclough 2003) and through the use of this discourse establish membership of a trans-national group. According to Gee (1996), Discourse with a capital "D"

> is a socially accepted association among ways of using language, other symbolic expressions, and 'artifacts' of thinking, feeling, believing, valuing and acting that can be used to identify oneself as a member of a socially meaningful group or "social network," or to signal (that one is playing) a socially meaningful "role." (p. 131)

Although I choose not to capitalise discourse, the definition of global discourse that I use in this chapter is in alignment with Gee (1996). Discussions about mathematics education that use shared terms and expressions to describe, for example, how mathematics should be taught and learnt, also situate the speakers as belonging to the community of mathematics educators. When mathematics educators go elsewhere in the world and talk about events, then they not only use language to shape how those events come to be seen as mathematics education, but they also identify themselves as members of the mathematics education community.

The differences, between earlier cases of global discourses affecting mathematics education and the present day situation, is that the global discourses that currently affect mathematics education often arise outside of mathematics education or even general education. For example, at the beginning of the twentieth century, it was educationalists, such as Dewey (Tyler 1987) and Montessori (1912), who spread the notion that children's interests should form the basis of their learning. This was in direct contrast to previous views that children should dislike what they were learning in order to develop discipline while being educated (Tyler 1987). Arithmetic, alongside algebra and geometry was considered as needed for developing this discipline (Klein 2003). In contrast, Cuban (2007) illustrated how the No Child Left Behind policy of US president, George W. Bush, which resulted in significant changes to how mathematics, among other subjects, was valued in schools, was a result of agitation by businesses wanting "nimble college-educated workforce for early 21st century labor markets" (p. 7). The propagators of global discourses, who seek to change how mathematics education events can be described, may originate with groups such as politicians or economists who now dominate curricula discussion (see for example, Lange and Meaney 2017, this volume).

Atweh et al. (2003) stated that, in mathematics education, globalisation could be seen in "the convergence of school mathematics and mathematics education curricula around the world" (p. 188). Nevertheless, global discourses should not be considered homogenous or in alignment with each other (Atweh and Clarkson

2002b) and, thus, convergence of curriculum does not have to be the end result of the integration of different global discourses. As well, what may look like convergence could hide important differences when policies are implemented. For example, colonisation in the nineteenth century resulted in Western curriculum, with connections to mathematics, being used in geographical spaces, a long way from where the curricula originated (Kamens and Benavot 1991). The circumstances of the spaces where the imported curricula arrived affected its implementation. In a later example, Blakers (1978) discussed how overseas influences affected mathematics curriculum development in Australia in a haphazard way, in the 1950s–1970s, partly because of the particularly circumstances that surrounded school education there. Although curricula may look similar, other factors such as the number of hours students attend school will affect how the curricula is implemented (Kamens and Benavot 1991).

Education policies, including curricula, may combine, but not necessarily integrate, contradictory discourses leaving mismatches and contradictions to be worked through and implemented by practioners (Otterstad and Braathe 2016; Apple 1995). As Angus (2004) stated "educational change is concerned with the negotiation and contestation of educational meaning and educational politics" (p. 26). The differences in philosophies expressed by prominent global discourses may negate some perspectives, not just at the practioner level, but also at the societal level. I explore these disjunctions in the section on Bernstein's pedagogic device.

Using examples from mathematics curricula from around the world, I explore how two global discourses, an equity discourse and a neo-liberal discourse, are combined, using Bernstein's (2000) pedagogic device. The pedagogic device provides ways to investigate the negotiation and contestation of knowledge transformation at different levels. Before discussing how the pedagogic device works, I consider mathematics curriculum development.

2 Mathematics Curricula and Their Relationship to Discourses

Mathematics curricula can be considered barometers, which reflect societal views about the role of mathematics in education (see for example, Smith and Morgan 2016; Wake and Burkhardt 2013). When different aspects of mathematics become valued as a result of changes in discourse, curricula are adapted to match these new values (Romberg 1993). The need for education to meet labour force requirements has re-emerged as a global discourse across the world. In relationship to mathematics education, this global discourse is often reflected into policies, as a need for more students to take up science, technology, engineering and mathematics (STEM) careers. For example, Wake and Burkhardt (2013) investigated the European Union's (EU) initiatives to introduce inquiry-based learning (IBL) in mathematics and science curricula. These initiatives were directly linked to ensuring that more students gained the necessary knowledge and skills for STEM careers. The EU's decision combined the global discourses about the need for more STEM graduates for the economic well-being of society, from outside the education field, and IBL as an appropriate method for teaching mathematics and science, from inside the education field. Although the EU's policies situate these global discourses as being in alignment, unless other aspects of the schooling system, such as assessment, also support the use of IBL, it is unlikely that IBL will retain its position of importance in school curricula of EU nation-states (Wake and Burkhardt 2013). In contrast, the global discourse about the need for more students to take up STEM careers is unlikely to be challenge because of implementation issues, as it forms an aim for curricula, that stands outside of attempts to achieve it.

Arguing for the need for a sociology of education, Bernstein (1971) stated "curriculum defines what counts as valid knowledge, pedagogy defines what counts as valid transmission of knowledge and evaluation defines what counts as valid realization of this knowledge on the part of the taught" (p. 47). As such, curricula are seen as the guiding documents for teachers to plan from, in that they contain "educational philosophy statements and general goals and, to varying degrees, the specific objectives, learning activities, teaching strategies, and assessment procedures" (Sirotnik 1988 p. 56). However, the transmission of knowledge through pedagogies, such as IBL, and assessment, such as rigorous testing regimes, affect, as well as being affected by, curricula (Cuban 2007). As well as noted by Cuban (2007), although curricula, pedagogy and assessment may change, historical (global) discourses remain in evidence, often resulting in hybrid policies and classroom practices.

Although the importance of curricula has been recognised as connected to "the distribution of power and principles of social control" (Bernstein 1971, p. 47), curriculum development is rarely the focus of mathematics education research, particular research with a sociopolitical focus (McMurchy-Pilkington et al. 2013). Curriculum development is acknowledged as a political activity, because of the ideologies held by the planners (Walker 1971), regardless of whether or not those involved in the planning are aware of them. In contrast, Walker's acknowledgement of the political aspect of curriculum development is incorporated into the first stage of his naturalistic model of curriculum:

> The system of beliefs and values that the curriculum developer brings to his task and that guide the development of the curriculum is what I call the curriculum's platform. The word "platform" is meant to suggest both a political platform and something to stand on. (Walker 1971, p. 52)

In the second stage, the *deliberation* stage, the developers interact and in so doing "defend their own platform statements and push "spur of the moment" ideas" (Print 1993, p. 76). The final stage is that of the design which links the product, the curriculum, to the decisions made in its production, whether they were implicit or explicit. An example of how the beliefs and values of the curriculum developers affect their decisions is that of how the teacher is situated in a curriculum. Autio (2014) stated:

> Bluntly put, is the teacher implicitly or consciously defined as a passive agent in the system (what s/he never is in reality!); an assumed conduit for external administrative, political and scientific ideas disciplines and mandates like in "implementation" policies; or an academically educated intellectual whose most significant work is trusted, supported or encouraged by surrounding culture and society? (p. 19)

Those who have the possibility to constitute what is valued in society use curriculum planning to determine the kinds of mathematics education social events that should be provided. Distributing different kinds of mathematics to different groups of children will affect the possibilities for developing thinking and, therefore, reinforce social control (Kollosche 2014) and the acceptance of global discourses.

The global discourse about the need for schooling to meet labour force requirements has a long history, which to some extent explains why it gained so much currency at the turn of the century. This global discourse reinforces the view that different types of children should receive different types of education, including mathematics education, to meet their anticipated differentiated, labour force requirements. For example, in England, in the nineteenth century "elementary schools were designed to produce a labour force" (Lawton 1984, p. 1) that would predominantly be used in factory work and for children to come "to be obedient and to have respect for the property of their betters" (Lawton 1984, p. 2), also necessary for factory work. Primary schools were funded by factory owners or other affluent people who controlled what was taught. In comparison, grammar and public schools provided an education in leadership which was done through "character training" and provided students with "the kind of knowledge which would be an obvious mark of their exclusive rank" (Lawton 1984, p. 1). Public schools were funded by wealthy parents, while grammar schools were funded by the government. The government employed their graduates as civil servants. The mathematics required for work in bureaucracy is also based on discipline and an acceptance of rules but includes a requirement to think logically (Kollosche 2014).

The mathematics available in each of these kinds of schools reflected the knowledge that society anticipated that these particular sets of students needed. As mass education was instigated across the world, arithmetic was at the vanguard of what young people were considered to need to learn at school (Kamens and Benavot 1991). Only some students received instruction in other topics of mathematics as there were far fewer jobs which required logical discipline, it was supposed to provide.

These distinctions continued into the twentieth century, but were accompanied by different sets of reasoning, through changes in discourses. Goodson (1989, p. 19) stated that the differentiated curricula in Britain suggested by the Norwood Report in 1949 were based on assessments of mental capabilities. Klein (2003) provided evidence of similar discussions in USA from the same period. A global discourse had developed around mental aptitude as delineating students' capacities to learn. This discourse replaced the discourse about the need to fulfil labour force requirements and became the new reason for providing groups of students with different curricula, even though the end result was that these groups were still channelled into different work futures.

However, at least from the 1970s, this differentiating of students at school by their performance in mathematics was resisted and attempts made to change curricula, see for example the work done by Freudenthal on this (Gravemeijer and Terwel 2000). In recent years, this resistance along with other realisations of how mathematics education contributed to inequitable outcomes for different groups of students has led to sustained calls for equity and access in mathematics education.

2.1   The Incorporation of Global Discourse About Equity and Access into Curricula

Discussions about equity in mathematics education have a long history (Secada 1989). For example, Secada (1989) referred to US legal cases about school segregation in the 1950s as highlighting concerns about access to educational opportunities provided by achievement in mathematics education. These concerns focused on low achievement and high rates of early drop-outs from school. These concerns formed the basis for the first version of the global discourse about equity, which was evident also in other parts of the world (see Román et al. 2015). At this time, Secada (1989) considered equity issues to be "about whether or not a given state of affairs is just" (p. 642).

In the 1990s, the global discourse about equity gained strength, although it was still predominantly used by mathematics education researchers who researched social justice issues (Meaney 2000). At this time, the concerns about the marginalisation of some groups from possibilities to gain good results in mathematics led to calls for curricula reform. The global discourse about problems with differentiating learning opportunities based on student results was combined with another global discourse, the need for quality mathematics education in schools. For example, Schoenfeld (2002) considered high quality curriculum as one of four conditions for ensuring that all students received high quality mathematics instruction. The need for "all" students to succeed in mathematics became a refrain connected to this global discourse (Pais 2014).

In the last quarter of a century, the global discourses about equity, problems with differentiating learning opportunities and the need for high quality learning opportunities, has contributed to many countries including rhetoric about access and quality in their curricula. An example can be seen in the National Council of Teachers of Mathematics' (2000) *Standards*:

> Creating, supporting, and sustaining a culture of access and equity require being responsive to students' backgrounds, experiences, cultural perspectives, traditions, and knowledge when designing and implementing a mathematics program and assessing its effectiveness. Acknowledging and addressing factors that contribute to differential outcomes among groups of students are critical to ensuring that all students routinely have opportunities to experience high-quality mathematics instruction, learn challenging mathematics content,

and receive the support necessary to be successful. Addressing equity and access includes both ensuring that all students attain mathematics proficiency and increasing the numbers of students from all racial, ethnic, linguistic, gender, and socioeconomic groups who attain the highest levels of mathematics achievement.

(http://www.nctm.org/Standards-and-Positions/Position-Statements/Access-and-Equity-in-Mathematics-Education/)

Although the NCTM standards are not endorsed by US state governments, they do provide inspiration for much mathematics teaching both in USA and elsewhere. As such, the standards (NCTM 2000) draw on global discourses but also reinforce these discourses and, thus, reinforce the necessity to use these global discourses to be part of the community of mathematics educators. This is evident in the number of articles that refer to this document when highlighting equity issues in mathematics education research, but also would include the informal discussions between mathematics teachers. For example, in critiquing calls for there to be more mathematics in mathematics education research, Martin et al. (2010) asked instead whether there should be more focus on equity in mathematics education research given that it was the first principle of the NCTM's standards.

In some curricula, the recognition of the need to provide more equitable access to mathematics learning experiences is implicit and situated in a general overview rather than in relationship to mathematics learning. This is the case in the Swedish curricula (Helenius 2015, personal communication). At other times, aspects of the equity discourse appear explicitly in curricula. For example, in the draft mathematics curriculum of Nepal, much of which was incorporated into the present curriculum (Bilbeck 2017, personal communication), there is a specific section on equity and inclusion in the key principles for reform:

The principle of equity and inclusion puts emphasis on a) making powerful mathematical ideas accessible to all children; and b) promoting appropriate pedagogical approaches to involve children in all possible mathematical learning endeavours. The notion of inclusion refers to increasing the participation of all learners in mathematics, thus reducing the possibility of their exclusion from classroom activities. Present educational practices seem to have been guided by the equality principal where learners are taken as having the same type of learning ability or similar learning styles. All the blame for low achievement goes to the learner's lack of aptitude. Impact of social background, suppression from the teachers, parents and curriculum and oppression of privileged group of people in society is not taken into account. So a democratic mechanism that promotes self respect of learners and encouragement for learning is to be developed. (Shrestha et al. 2012, p. 33)

This call for equity is closely related to School Sector Development Plan (Government of Nepal 2016) which has equity as the first of its five dimensions, driving societal improvement through education.

It could be expected that with clearly stated goals for equity and increased access through curricula more equitable outcomes for different groups of students could be achieved. Certainly Schoenfeld (2002) suggested that there were signs that the gap between different groups of students' mathematics achievement was flattening out as a consequence of US states aligning their curricula with the NCTM standards. However, particularly in the last ten years it would seem that high quality mathematics learning opportunities have not become more accessible. The Organisation of Economic Development (OECD) (2013) stated that the 2012 mathematical literacy of 15 year old students compared to 2003 had improved only slightly in regard to how much socio-economic status could predict achievement (down from 17 to 15%). Only in three countries, Turkey, Mexico and Germany, did mathematics achievement improve alongside a reduction in the prediction value of disadvantage. The OECD indicated that this showed that improvement in a country's mathematics achievement does not have to be at the cost of equity. However their comparison of results across 10 years does not indicate that this kind of improvement was particularly simple to achieve if only 3 countries were able to produce it. It may be that the inclusion of contradictory discourses results in relevant research evidence on improving access and quality of mathematical learning opportunities being ignored:

> Even as policy texts express concern about the risk of social exclusion, there is continued reliance on restricted forms of evidence, on performance measurement and management, and on superficial and contradictory acknowledgements of difference and diversity. As a consequence there is a failure to take full account of social science research-based evidence that is relevant to meeting the challenges posed by such risky, complex and unjust contexts. (Ozga and Jones 2006, p. 3)

Ozga and Jones (2006) indicate that even when calls for improvements in equity are made, through curricula, then these improvements may be counter-balanced by being judged through inappropriate measures, such as the OECD's testing regime, and by using descriptions of diversity, such as socio-economic status that are fraught with problems (see Valero and Meaney 2014). Pais (2014) went further in his critique by stating that "mathematics for all" can never be achieved because of the system's reliance on some students failing, in order to fulfil capitalist requirement that education systems provide the necessary credits to only some students, to gain high-paying jobs.

Consequently, calls for equity may be ineffectual, particularly when this global discourse is in conflict with other global discourses. This affects how the social events of mathematics education can be described—only in terms of success in tests —so that members of the mathematics education community are unable to have shared perspectives about how the aims for equity can be integrated in meaningful ways into those social events (Meaney 2000).

2.2 Neo-liberal Discourses and Their Effect on Mathematics Education

A global discourse that has a direct effect on the inclusion of equity into mathematics education social events, is that of neo-liberalism. Originating in economics, the neo-liberal global discourse affects education because "technologies of institutional control and accountability are identified as the key mechanisms which now increasingly subordinate education to the economic imperatives of the latest phase of globalised, 'post-industrial' capitalism" (Beck 1999, p. 227). Lingard (2010) suggested that the neo-liberal discourse manifests itself in common sense discussions about competition between schools, coupled with parental pressure and the need for parents to choose schools for their children. Competition, as determined by test results, is considered as naturally leading to improved education (Llewellyn 2017). As such, neo-liberal discourses are detrimental to instigating access to high quality mathematics learning opportunities for all students, regardless of class or ethnicity. If all students had access to high quality learning opportunities, then there would be no need for competition to drive improvement.

For example, Sweden's results in international testing declined over a decade, with the proportion of students at the lowest levels of performance increasing (OECD 2015). Although the latest report suggests that mathematics results have begun to improve, the difference between students living in high socio-economic areas and those in low socio-economic areas continues to increase (OECD 2016). In Sweden, it would seem that for the equitable outcomes outlined in the curricula cannot be achieved when school choice is in operation (Östh et al. 2013). As Östh et al. (2013) found:

> With expanding school choice, the differences between schools have increased and, at the same time, Sweden's comparative performance has declined (OECD 2010). Thus, as has been the case with other neoliberal ideas, school choice—when tested— has not been able to deliver the results promised by theoretical speculation. (p. 422)

In Australia, the gap in achievement between those attending schools in high socio-economic areas and those attending schools in low-socioeconomic areas has also increased, particularly in Year 9 numeracy results (Bonnor and Shepherd 2014). Similar suggestions to those made in Sweden about the causes of the problem, parental school choice, have been raised (McConney and Perry 2010).

As well as parental school choice, high-stakes assessments are recognised as having a significant impact on what is taught within education systems (Lange and Meaney 2012). The deep learning of mathematics advocated in equity statements is often replaced by a focus on test content, particular for students in schools where test results have wider implications for teachers' jobs (Lange and Meaney 2012). For example, an evaluation of mathematics teaching in Nepal suggested that "no matter how the syllabus and its objectives are genuine and child-centred, it would be a tough challenge for teachers to change their teaching under the current examination system" (Nakawa 2013, p. 131).

Lingard (2010) described the likely impact of the introduction on high-stakes national testing in Australia as "what we will most likely see is test-focused schooling, with a consequent narrowing of curricula and pedagogies, with this having its most egregious effects in low SES schools" (p. 131). Children attending schools in low socio-economic areas were likely to have learning focused on passing the numeracy and literacy tests and a reduction in learning opportunities in other subject areas (Taylor et al. 2003).

Although Lingard (2010) suggested that educational accountability can be achieved in other ways that better recognise research evidence and specific community needs, it is unlikely that this is a realistic option. This is because, although it originated outside of education, the neo-liberal discourse connects more strongly with historical views of the role of mathematics education in schools. It has therefore been able to adapt existing discourses about the necessity to distribute different kinds of knowledge to different groups of students. In the next section, I use Bernstein's (2000) pedagogic device to suggest how the operation of neo-liberal discourse at the different levels of the pedagogic device overtakes and disrupts equity discourses. This results, not in a hybrid of the two discourses (Cuban 2007), but a parasitic invasion of one discourse by the other.

3 The Pedagogic Device

Bernstein (2000) used the notion of the pedagogic device to explain the "social grammar" that reproduces and transforms knowledge within education systems, often invisibly to the detriment of those from low socio-economic backgrounds. The pedagogic device has been used by Kanes et al. (2014) to describe how those who instigate OECD's testing regime have come to regulate how students relate to mathematics. Bernstein stated "those who own the device own the means of perpetuating their power through discursive means and establishing, or attempting to establish, their own ideological representations" (p. 114). Knowledge is transformed through a hierarchical set of rules:

> *Distributive rules*: These rules distributed forms of knowledge to different social groups. In this way, distributed rules distributed different forms of consciousness to different groups. Distributive rules distributed access to the 'unthinkable', that is, the possibility of new knowledge, and access to the 'thinkable', that is, to official knowledge.
>
> *Recontextualising rules*: These rules constructed the 'thinkable', official knowledge. They constructed pedagogic discourse: The 'what' and the 'how' of that discourse.
>
> *Evaluative rules*: These rules constructed pedagogic practice by providing the criteria to be transmitted and acquired. (Bernstein 2000, p. 114)

At each point of transformation, there is contestation of what knowledge becomes acceptable. As different groups come to value different knowledge, through the adoption and adaption of global discourses, then what is contested will change as will the knowledge, which comes to be considered acceptable.

By focusing on the knowledge which is distributed, recontextualised and evaluated into mathematics curricular, I illustrate how it is regulated by global discourses, including those about equity and neo-liberalism. These discourses both determine what is valuable and then enshrine this valuation by reinforcing through curricula the need for all members of the mathematics education community to discuss mathematics education in this way. This approach is in alignment with Loughland   and   Sriprakash   (2016)   who   used   "Bernstein's   ideas   of recontextualisation to show how discourses of competition, standardisation and commensurability become 'inside' discourses of current Australian education policy ambitions relating to social and educational equity" (p. 234). Their analysis focused on the adaptation of neo-liberalism as a discourse about economic markets to one about education, so that existing structural inequalities in the Australian education system were ignored and the equity discourse became shaped into one about the need for assessment and parental choice, based on reports about school performance. Although such adaptations are also present in the specific case of mathematics education, the historical precedence of mathematics curricula being differentiated for specific groups of students and of testing to be used to determine this differentiation provide additional information about how Bernstein's pedagogic device operates. The analysis of the mechanism of how knowledge is valued and transformed at different levels of the pedagogic device illustrates how the ascendancy of the neo-liberal discourses and the disregarding of the discourse about more accessible, high-quality mathematics learning opportunities has been achieved.

3.1 Distribution Rules

The distribution rules of the pedagogic device allow for alternative scenarios to become possible, for example within curricula, because of the discursive gap resulting from differences between perceptions of what is considered valuable (Au 2008). This gap provided an opportunity for the insertion of the equity discourse into mathematics curricula, through the recognition of such things as deep learning being important for all students. Nevertheless, those who control the distribution rules are the ones who decide what should be included. As Au (2008) stated, "the distributive rules also seek to regulate not only what is thought of as possible or impossible, but also who has the right or power to set the limits of possibility" (p. 642).

Although Bernstein (2000) conceded that new knowledge could be determined outside of traditional fields, it is generally up to the proponents of the field to recognise and incorporate knowledge as valuable into the field. For example, both Jahnke (2012) and McMurchy-Pilkington et al. (2013) describe curriculum development processes which included consultations with teachers and teacher educators, but where the final version of curricula had to be sanctioned by the Ministry of Educations in their respective countries, Sweden and New Zealand. In Romania, Singer (2008) described the case of curriculum developers ignoring information from international tests showing Romania's general poor performance, while maintaining their belief in their students being high achievers due to their performance in mathematics Olympiads. This contributed to the failure of curriculum reform, initiated by the World Bank. Thus, those in charge of curriculum development control the distribution of knowledge into curricula and their adoption of specific discourses support their decision making about what is the valuable knowledge to be included.

As noted earlier, often historical understandings about what is valuable knowledge is merely adapted to suit the emphases of new global discourses. In documenting the dissatisfaction with the "New Math" movement of the late 1950s–1960s, Fey (1978) claimed that there was a call for more practical mathematics, which had been the basis of many mathematics courses taught previously. The mathematics courses, which replaced the "New Math", were not the same as had been provided earlier. Nevertheless, the established ways of describing mathematics education allowed for an acceptable replacement to be instigated. Fey also stated "much of the challenge to make school programs more practical and accountable for their effectiveness seems to reflect anxiety about personal economic pressures much more than philosophical disagreement about educational policy" (p. 352). Thus, historically accepted ways of discussing mathematics education are adapted to suit local and particular circumstances.

The discourse around equity and access has not had an established role in the field of mathematics curriculum development, whereas discussions about assessment that can be linked to neo-liberal discourse have always been part of this field. For example, Howson (1978) documented the case on a school inspector in 1859 who asked 1344 children two questions, one of which was on arithmetic. On the basis of the answers, this inspector categorised 53 schools as good, fair or inferior. I contend that this historical backdrop to mathematics curriculum development has contributed to the discussion of equity being transformed into one about raising standards and the need for skills for work, which are in alignment with neo-liberal discourses about the priority of economic interests (Lingard 2010). By focusing on economic needs, the inherent inequities within the schooling system are disregarded and the need for improving educational outcomes becomes the responsibility of individual learners.

3.2 Recontextualising Rules

The recontextualising rules take the knowledge identified as valuable by the distribution rules and places it within a pedagogic discourse (Au 2008). The resulting pedagogic discourse provides information not just on the content that should be taught but also on how it should be taught. As Au (2008) stated "the recontextualization of knowledge into pedagogic discourse is ultimately connected to external socio-economic relations that grant teachers, schools, districts, and governing bodies the power to make decisions regarding the content and form of knowledge" (p. 642). In making these decisions, these organisations will draw on global discourses to situate themselves as appropriate members of the groups who can make such decisions.

Morgan and Xu (2011) outlined a number of influences on how the recontextualising rules operate in China and UK in relationship to the pedagogic discourse around mathematics education. Bernstein (2000) identified two components of the recontextualising field, the official recontextualising field controlled by the state, and the pedagogical recontextualising field, influenced by textbook writers and teacher educators, outside of the direct control of the state. Morgan and Xu (2011) described these sub-fields as being more integrated in China as the state controlled textbook writing and teacher education, in more overt ways than in the UK.

In contrast to the situation described by Morgan and Xu (2011), I contend that the instigation of surveillance mechanisms to make teacher education and schools accountable, reduce the independence of the pedagogical recontextualising field, available to teachers and teacher educators. In the first half of the twentieth century, teacher educators in teacher education colleges took over much of the responsibility for determining mathematics curricula (see as a description of the US situation Klein 2003). This control has been reduced over time and rather than being seen as leaders in the field, they are also situated as needing to be regulated. Alongside schools in many countries being regularly inspected (see for example, Jahnke 2012), teacher education institutions are also regulated through the need for them to gain accreditation. In Europe, there has been a recent increase in accreditation to meet European Union requirements for mutual recognition of tertiary qualifications (Tatto et al. 2012). Although accreditation is set up to ensure the quality of new teachers, it is also likely to reduce the possibilities for instilling equity aspects into mathematics teacher education. This is because mathematics teacher education often focuses on the mathematical knowledge (see for example, Meaney and Lange 2012). Tests of preservice teachers' mathematical pedagogical content knowledge rarely include explicit questions connected to dealing with equity issues in mathematics classrooms. For example, no such questions appear in Tatto et al. (2012).

It is likely that the newness of the equity global discourse in discussions about mathematics education means that it is under-recognised by those making decisions about how to recontextualise curricula as pedagogic discourse. In USA, Sleeter (2008) described how universities had only a limited period in the 1990s to incorporate equity issues across teacher education programmes before neo-liberal discourses began to take precedence. Consequently, it can be said that the recontexualising of the neo-liberal discourses emphasised amongst other things, content knowledge above professional knowledge. So, although curricula may indicate the need to consider issues of access and quality, teacher education programmes have limited opportunities to provide preservice teachers with understandings of how to do this.

3.3   Evaluative Rules

The evaluative rules determine what knowledge is reproduced by teachers to students in classrooms (Bernstein 2000). Classroom practices are enacted by teachers based on curricula recommendations. Even when curricula include a focus on equity, the enactment of it in classroom practices can be whittled down to an almost unrecognisable version. Loughland and Sriprakash (2016) showed how the evaluative rules complete the narrowing of understandings about equity to be "something that is achieved through market notions of competition and commensurability" (p. 240). Teachers may speak the discourse of equity, but enact it in alignment with notions from the neo-liberal global discourse. As a consequence, social events of mathematics education cannot lead to an overcoming of injustice.

As noted earlier, national tests can determine how mathematics learning should be realised and in so doing set out for teachers, students and the rest of the community what constitutes mathematics. In an analysis of PISA test items, Kanes et al. (2014) showed that although students seemed to be provided with an opportunity to show deep learning on the mathematics to do with climate change, a topic with real-world implications, this was illusionary. Instead contradictory information was provided about how the student should situate themselves in regard to answering the question. Kanes et al. (2014) suggested that this was likely to result in students, who had the skills for manoeuvring through the contradictions, becoming self-governing, rather than critical thinkers.

Lange and Meaney's (2014) research on public discourse about national tests in Australia showed a general acceptance that what was in the numeracy tests was the important mathematics that students needed to know. These views by the general public, politicians and others, are likely to have been built on historical understandings that while all children should learn arithmetic, only some can learn mathematics (Howson 1974). Similarly, Morgan and Xu (2011) identified such views in interviews with Chinese teachers. Wake and Burkhardt (2013) concluded in regard to teachers' resistance to changing practices to meet curricula expectations, "it is often the well- and long-established expectations of the community that provide obstacles to policy intentions being realized" (p. 853). Curricula that include requirements to consider access and quality issues in mathematics teaching are unlikely to be implemented in a broad and meaningful way, when mathematics is what is needed for labour force requirements, but differentiated so that students' work careers are determined by what they cannot achieve in mathematics tests.

4 Implications

In this paper, I argue that although equity issues as a global discourse have become included within mathematics curricula across the world, the neo-liberal discourse has controlled how this equity discourse came to be realised in classroom practices. This is in alignment with Angus (2004), who in a micro-political, ethnographic study, showed how actors within educational systems come to be complicit in accepting neo-liberal norms.

In regard to mathematics education, neo-liberal discourse has the ability to tap into historical discourses amongst other things, mathematics and assessment and mathematics and students' ability. This is reinforced by government surveillance, both of schools and teacher education institutions, which emphasise content learning over professional knowledge. Through the neo-liberal global discourse, students are situated, by educators, parents and the wider society, as responsible for their own learning. Consequently equity discourses about access and quality, so that educational outcomes are more just, are reduced to mere considerations about the mathematics students need to pass in order to fulfil particular work requirements or to access further study. Accountability through regular assessment becomes the only way of judging whether students have gained access to quality mathematics education. Their results, rather than being seen as judgements on how the system is providing equitable opportunities, become about students being self-regulated learners. When students do not achieve, they become the owners of their own failure.

As can be seen in many of the contributions to this volume, mathematics education researchers who are interested in how inequality, power, access and identify are integrated into mathematics education tend to focus either at the classroom level or at the societal level. Yet it is essential to see these two levels are seen as knitted together, where each level both draws on and reinforces the other. Curricula investigations provide the necessary opportunities for exploring how discourses integrated into societal understandings about mathematics education become classroom episodes that limit students' possibilities to receive high-quality learning opportunities. In this chapter, I have begun a discussion of how curricula investigation can illustrate how the well-intentioned calls for the inclusion of equity understandings be included into curricula become subsumed into neo-liberal discourses. However, much more research is needed if the complex interplay between classroom interactions and global discourses are to be understood.

Although Bernstein (1990) indicated that the pedagogic device maintains class distinctions through education, he also identified that as knowledge is selected through the operation of the different rules of the device, a gap between abstract meanings and immediate contexts can appear. As Au (2008) discussed, this gap contradicts the regulation of thought, which restricts what knowledge is considered valuable, by enabling people to become aware of the 'unthinkable'. As global discourses of neo-liberalism and social equity are distributed, recontextualised and evaluated through curricula into local mathematics education social events, previous unthinkable possibilities can be made available. The challenge is to use these possibilities so that the equity discourse is given more than cursory attention when mathematics curricula are implemented and not reduced merely to discussion of performance indicators. However, opportunities to make use of these possibilities will close if not acted upon. It may be that the possibility offered by the equity global discourse can act as a mirror through which a willingness to conform to changes in alignment with the neo-liberal discourse can be challenged.

References

Angus, L. (2004). Globalization and educational change: Bringing about the reshaping and re-norming of practice. *Journal of Education Policy, 19*(1), 23–41.
Apple, M. W. (1995). *Education and power*. New York: Routledge.

Atweh, B., & Clarkson, P. (2002a). Globalized curriculum or global approach to curriculum reform in mathematics education. *Asia Pacific Education Review, 3*(2), 160–167.

Atweh, B., & Clarkson, P. (2002b). Globalisation and mathematics education: From above and below. In *Problematic Futures: Educational Research in an Era of Uncertainty: Proceeding of the Australian Association of Research in Education Conference. Conference held December 1–5, 2002.* University of Queensland. Available from: http://www.aare.edu.au/publications-database.php/3372/globalisation-and-mathematics-education-from-above-and-below.

Atweh, B., Clarkson, P., & Nebres, B. (2003). Mathematics education in international and global contexts. In A. J. Bishop, M. A. Clements, C. Keitel, J. Kilpatrick, & F. K. S. Leung (Eds.), *Second international handbook of mathematics education* (pp. 185–229). Dordrecht: Springer.

Au, W. W. (2008). Devising inequality: A Bernsteinian analysis of high-stakes testing and social reproduction in education. *British Journal of Sociology of Education, 29*(6), 639–651.

Autio, T. (2014). Internationalization of curriculum research. In W. F. Pinar (Ed.), *International handbook of curriculum research* (pp. 17–31). New York: Routledge.

Beck, J. (1999). Makeover or takeover? The strange death of educational autonomy in neo-liberal England. *British Journal of Sociology of Education, 20*(2), 223–238.

Bernstein, B. B. (1971). On the classification and framing of educational knowledge. In M. F. D. Young (Ed.), *Knowledge and control* (pp. 47–69). London: Collier-Macmillan Publishers.

Bernstein, B. (1990). *The structuring of pedagogic discourse.* London: Routledge.

Bernstein, B. (2000). *Pedagogy, symbolic control and identity: Theory, research, critique* (rev ed.). Lanham, MD: Rowman & Littlefield Publishers.

Blakers, A. L. (1978). Change in mathematics education since the late 1950's-ideas and realisation Australia. *Educational Studies in Mathematics, 9*(2), 147–158.

Bonnor, C., & Shepherd, B. (2014). *School equity since Gonski: Since bad became worse.* Unknown: Need to succeed alliance. Available from: http://needtosucceed.org/wp-content/uploads/2014/09/School-equity-since-Gonski-1.pdf.

Cuban, L. (2007). Hugging in the middle. Teaching in an era of testing and accountability, 1980–2005. *Education Policy Analysis Archive, 15*(1). Available from http://epaa.asu.edu/epaa/v15n1/.

Fairclough, N. (2003). *Analysing discourse: Textual analysis for social research.* London, UK: Routledge.

Fey, J. T. (1978). Change in mathematics education since the late 1950's—Ideas and realisation USA. *Educational Studies in Mathematics, 9*(3), 339–353.

Gee, J. (1996). *Social linguistics and literacies: Ideology in discourses* (2nd ed.). Bristol, PA: Taylor & Francis.

Goodson, I. F. (1989). "Chariots of Fire": Etymologies, epistemologies and the emergence of curriculum. In G. Milburn, I. F. Goodson, & R. J. Clark (Eds.), *Re-interpreting curriculum research: Images and arguments* (pp. 13–25). London: Falmer Press.

Government of Nepal. (2016). *School sector development plan 2016–2023.* Kathmandu: Ministry of Education, Government of Nepal.

Gravemeijer, K., & Terwel, J. (2000). Hans Freudenthal: A mathematician on didactics and curriculum theory. *Journal of curriculum studies, 32*(6), 777–796.

Howson, G. (1974). Mathematics: The fight for recognition. *Mathematics in School, 3*(6), 7–9.

Howson, A. G. (1978). Change in mathematics education since the late 1950's-ideas and realisation Great Britain. *Educational Studies in Mathematics, 9*(2), 183–223.

Jahnke, A. (2012). *The process of developing a syllabus: Reflections of a syllabus developer.* Paper to be delivered at the 12th International Congress of Mathematics Education, Seoul, July 8–15, 2012. Available from: http://www.icme12.org/sub/tsg/tsg_last_view.asp?tsg_param=32.

Kamens, D. H., & Benavot, A. (1991). Elite knowledge for the masses: The origins and spread of mathematics and science education in national curricula. *American Journal of Education, 99*(2), 137–180.

Kanes, C., Morgan, C., & Tsatsaroni, A. (2014). The PISA mathematics regime: Knowledge structures and practices of the self. *Educational Studies in Mathematics, 87*(2), 145–165.

Klein, D. (2003). A brief history of American K–12 mathematics education in the 20th century. In J. Royer (Ed.), *Mathematical cognition: A volume in current perspectives on cognition, learning, and instruction* (pp. 175–225). Charlotte, NC: Information Age.

Kollosche, D. (2014). Mathematics and power: An alliance in the foundations of mathematics and its teaching. *ZDM Mathematics Education, 46*(7), 1061–1072.

Lange, T., & Meaney, T. (2012). The tail wagging the dog? The effect of national testing on teachers' agency. In C. Bergsten, E. Jablonka, & M. R. Sundström (Eds.), *Evaluation and comparison of mathematical achievement: Dimensions and perspectives: Proceedings from MADIF 8* (pp. 131–140). Linköping: Svensk Förening för Matmematikdidaktisk Forskning.

Lange, T., & Meaney, T. (2014). It's just as well kids don't vote: The positioning of children through public discourse around national testing. *Mathematics Education Research Journal, 26*(2), 377–397.

Lange, T., & Meaney, T. (2017). The production of "common sense" in the media about more mathematics in early childhood education. In M. Jurdak & R. Vithal (Eds.), *Social and political dimensions of mathematics education*. New York: Springer.

Lawton, D. (1984). Curriculum and culture. In M. Skilbeck (Ed.), *Readings in school-based curriculum development* (pp. 275–289). London: Paul Chapman.

Lingard, B. (2010). Policy borrowing, policy learning: Testing times in Australian schooling. *Critical Studies in Education, 51*(2), 129–147.

Llewellyn, A. (2017). Technologies of (re)production in mathematics education research: Performance of progress. In H. Straehler-Pohl, N. Bohlman, & A. Pais (Eds.), *The disorder of mathematics education: Challenging the socio-political dimensions of research* (pp. 153–169). New York: Springer.

Loughland, T., & Sriprakash, A. (2016). Bernstein revisited: The recontextualisation of equity in contemporary Australian school education. *British Journal of Sociology of Education, 37*(2), 230–247.

McConney, A., & Perry, L. B. (2010). Science and mathematics achievement in Australia: The role of school socioeconomic composition in educational equity and effectiveness. *International Journal of Science and Mathematics Education, 8*(3), 429–452.

McMurchy-Pilkington, C., Trinick, T., & Meaney, T. (2013). Mathematics curriculum development and Indigenous language revitalisation: Contested spaces. *Mathematics Education Research Journal, 25*(3), 341–360.

Martin, D. B., Gholson, M. L., & Leonard, J. (2010). Mathematics as gatekeeper: Power and privilege in the production of power. *Journal of Urban Mathematics Education, 3*(2), 12–24. Available from: http://education.gsu.edu/JUME.

Meaney, T. (2000). *The process of valuing in mathematics education.* Paper presented at International Congress of Mathematics Education 9, Working Group—Social Justice in Mathematics Education, July 2000. Tokyo, Japan.

Meaney, T. (2014). Back to the future? Children living in poverty, early childhood centres and mathematics education. *ZDM Mathematics Education, 46*(7), 999–1011.

Meaney, T., & Lange, T. (2012). Knowing mathematics to be a teacher. *Mathematics Teacher Education and Development, 14*(2), 50–69.

Montessori, M. (1912). *The Montessori method* (A. E. George, Trans.). New York: Frederick A. Stokes.

Morgan, C., & Xu, G. R. (2011, July). *Reconceptualising 'obstacles' to teacher implementation of curriculum reform: Beyond beliefs.* Paper given at Mathematics Education and Contemporary Theory Conference. Manchester Metropolitan University, UK.

Nakawa, N. (2013). Current situations in pre-primary and primary mathematics in Kathmandu, Nepal. *Tokoyo Future University Research Bulletin, 6*, 121–139. Available from: http://www.tokyomirai.ac.jp/info/research/bulletin/06.html.

National Council of Teachers of Mathematics. (2000). *Principles and standards for school mathematics.* Reston, VA: NCTM.

OECD. (2010). *PISA 2009 results. Learning trends: Changes in student performance since 2000* (Vol. 5). Paris: OECD.

OECD. (2013). *PISA 2012 results: Excellence through equity: giving every student the chance to succeed* (Vol. II). Paris: OECD. https://doi.org/10.1787/9789264201132-en.

OECD. (2015). *Improving schools in Sweden: An OECD perspective*. Paris: OECD.

OECD. (2016). *Country note, Programme for International Student Assessment (PISA), results from PISA 2015: Sweden*. Paris: OECD. Available from: https://www.oecd.org/pisa/PISA-2015-Sweden.pdf.

Otterstad, A. M., & Braathe, H. J. (2016). Travelling inscriptions of neo-liberalism in Nordic early childhood: Repositioning professionals' for teaching and learnability. *Global Studies of Childhood, 6*(1), 80–97.

Ozga, J., & Jones, R. (2006). Travelling and embedded policy: The case of knowledge transfer. *Journal of Education Policy, 21*(1), 1–17.

Östh, J., Andersson, E., & Malmberg, B. (2013). School choice and increasing performance difference: A counterfactual approach. *Urban Studies, 50*(2), 407–425.

Pais, A. (2014). Economy: The absent centre of mathematics education. *ZDM Mathematics Education, 46*(7), 1085–1093.

Print, M. (1993). *Curriculum development and design*. Sydney: Allen & Unwin.

Román, H., Hallsén, S., Nordin, A., & Ringarp, J. (2015). Who governs the Swedish school? Local school policy research from a historical and transnational curriculum theory perspective. *Nordic Journal of Studies in Educational Policy, 2015*(1). Available from: http://www.tandfonline.com/doi/full/10.3402/nstep.v1.27009.

Romberg, T. A. (1993). How one comes to know: Models and theories of the learning of mathematics. In M. Niss (Ed.), *Investigations into assessment in mathematics education* (pp. 97–111). Dordrecht: Kluwer.

Schoenfeld, A. H. (2002). Making mathematics work for all children: Issues of standards, testing, and equity. *Educational Researcher, 31*(1), 13–25.

Secada, W. G. (1989). Agenda setting, enlightened self-interest, and equity in mathematics education. *Peabody Journal of Education, 66*(2), 22–56.

Shrestha, M. M., Tuladhar, B. M., & Koirala, S. P. (2012). *National framework for mathematics: Pres-school to grade 12 (proposed)*. Kathmandu: Council for Mathematics Education, Nepal Mathematics Society, Nepal Mathematics Centre.

Singer, M. (2008). Balancing globalisation and local identity in the reform of education in Romania. In B. Atweh, M. Borba, A. Barton, D. Clark, N. Gough, C. Keitel, C. Vistro-Yu, and R. Vithal (Eds.), *Internationalisation and Globalisation in Mathematics and Science Education* (pp. 365–382). Dordrecht: Springer.

Sirotnik, K. A. (1988). What goes on in classrooms? Is this the way we want it? In L. E. Beyer & M. W. Apple (Eds.), *The curriculum: Problems, politics and possibilities* (pp. 56–76). New York: State University of New York Press.

Sleeter, C. (2008). Equity, democracy, and neoliberal assaults on teacher education. *Teaching and Teacher Education, 24*(8), 1947–1957.

Smith, C., & Morgan, C. (2016). Curricular orientations to real-world contexts in mathematics. *The Curriculum Journal, 27*(1), 24–45.

Tatto, M. T., Peck, R., Schwille, J., Bankov, K., Senk, S. L., Rodriguez, M., … & Rowley, G. (2012). *Policy, practice, and readiness to teach primary and secondary mathematics in 17 Countries: Findings from the IEA teacher education and development study in mathematics (TEDS-M)*. Amsterdam: International Association for the Evaluation of Educational Achievement.

Taylor, G., Shepard, L., Kinner, F., & Rosenthal, J. (2003). *A survey of teachers' perspectives on high-stakes testing in Colorado: What gets taught, what gets lost*. Santa Cruz: University of California.

Tyler, R. W. (1987). The five most significant curriculum events in the twentieth century. *Educational Leadership, 44*(4), 36–38.

Wake, G. D., & Burkhardt, H. (2013). Understanding the European policy landscape and its impact on change in mathematics and science pedagogies. *ZDM Mathematics Education, 45* (6), 851–861.

Walker, D. F. (1971). A naturalistic model for curriculum development. *School Review, 80*(1), 51–65.
Valero, P., & Meaney, T. (2014). Trends in researching the socioeconomic influences on mathematical achievement. *ZDM Mathematics Education, 46*(7), 977–986.

Policy Production Through the Media: The Case of More Mathematics in Early Childhood Education

Troels Lange and Tamsin Meaney

Abstract This chapter explores how politicians' use of the media can disrupt educational traditions. Analysis of the discursive resources that a Norwegian Minister of Education used in a single authored debate article in a Norwegian newspaper shows that he drew on a well-known argument for why schools should teach mathematics, that of the need for socio-economic development of society. The use of this argument, rather than other arguments such as those about civic development, which would be more in alignment with the social pedagogy approach traditionally characterising early childhood education in Norway, seems to indicate that the Minister was promoting a shift in approach to one of preparing children for school. This example of the use of the media to determine how policy shifts are made is explored in relationship to promoting a new kind of "common sense" which does not require public discussion or input from mathematics education researchers.

Keywords Media rhetoric · Early childhood mathematics · Policy shifts
Economic development · Politicians

T. Lange (✉) · T. Meaney
Faculty of Education, Western Norway University of Applied Sciences,
Bergen Campus, Bergen, Norway
e-mail: troels.lange@hvl.no

T. Meaney
e-mail: tamsin.jillian.meaney@hvl.no

© Springer International Publishing AG 2018   191
M. Jurdak and R. Vithal (eds.), *Sociopolitical Dimensions of Mathematics Education*, ICME-13 Monographs, https://doi.org/10.1007/978-3-319-72610-6_11

1   Introduction

In this chapter, we explore a newspaper debate article, in which the Minister of Education in Norway attempts to present the shift in policy about the role of mathematics in *barnehage*[1] as "common sense", and in so doing limits public discussion of it. The study is part of a larger project about how policy documents and public discourse frame staff and parents' perceptions of mathematics education in barnehage. Our analysis considers how the structure of the debate article and the use of rhetorical devices, contribute to situating the Minister's argument as common sense. Although only one example, we see it as illustrating a trend in how politicians redefine mathematics education policy through determining how policy discussions are framed so that research outcomes become irrelevant, a practice connected to politicians' use of the media since at least the 1990s. Therefore, the case that we present in this chapter is part of a wider story of how politicians seek to change the ways that educational policy is introduced and discussed (Lingard and Rawolle 2004).

In the last two decades, there has been a growing realization that the media have a significant role in profoundly changing "social institutions and cultural processes" (Hjarvard 2008, p. 106). This includes using the media to affect education policy (Franklin 2004; Hattam et al. 2009; Lange and Meaney 2014; Stack 2006). For example, Hattam et al. (2009) showed how an Australian Minister for Education used the media, through a specific contact, to present school education as being in crisis and to blame teachers and their innovative pedagogies for this crisis. In this way, common sense understandings about educational issues are redefined to suit politicians' own interests. As Franklin (2004) noted about the Labour government in the UK:

> Politicians' preference for soundbites above sustained policy debate reflects the extent to which their determination to set the news agenda and to use media to inform, shape, and manage public discourse about policy and politics have become crucial components in a modern statecraft and system of governance. (p. 256)

Lingard and Rawolle (2004) discuss how politicians, in a variety of ways, use soundbites of short pieces of information about such things as new policy that is catchy and likely to stay in the memory of those who hear it without requiring an explanation. Soundbites are also used to ensure that the politicians have the final word in any discussion. The soundbite becomes the important point, reiterated in different media presentations. The release of these well-crafted pieces of information also replaces the need for the media and the public to engage in thorough

---

[1]The word *barnehage* (*barnehager* in plural) is the Norwegian term for institutions providing early childhood education and care for 1–5 year old children. Barnehage literally means a children garden. It is commonly translated to the German kindergarten (although not capitalised as in German). The organisation and naming of institutions for early childhood education and care varies significantly across countries. Hence, in order to maintain the situatedness of the study we have chosen to use the Norwegian term throughout the paper.

analysis of policy. Consequently, the media, rather than querying new policy, including educational policy, and providing opportunities to discuss it, can be manipulated by politicians to represent policy changes as an adjustment to match the general public's common sense understanding of the world.

The use of common sense to justify positions lies not just in the realm of politicians. Aspects of mathematics education are also often described in terms of common sense. For example, common sense has been defined by Radford (2008) as the things so taken-for-granted that they are not even noticed. Consequently, assumptions on which the common sense is based are also taken as pre-existing and given in any discussion. Keitel and Kilpatrick (2005) defined common sense as "a concept referring to local, situated or everyday knowledge" (p. 105). Therefore, common sense has come to be seen as knowledge based on everyday experience rather than knowledge based on rationality and logic. In mathematics education research, common sense has been use to describe attitudes towards mathematics (Gellert et al. 2001); the relationship between rankings in international comparative tests and a country's economic potential (Sjøberg 2015); and the learning of mathematics (Gravemeijer and Doorman 1999). Although common sense is recognised as affecting various aspects of mathematics education, little research has been conducted on how that common sense is produced or changed. In previous research (Lange and Meaney 2014), we showed how discussions in the media about national testing affected what mathematics in schools was considered to be, calculations and multiplication tables. This common sense acceptance about mathematics, which was not reflected in curricula documents, generally went unchallenged. LeBlanc (2012) investigated how Canadian media created a discussion of a report advocating the need for traditional methods of teaching mathematics through a series of rhetorical devices to present this view as being common sense.

Thus, there is recognition that the media among other institutions, including schools, contributes to the construction of common sense, by determining what can be challenged in discussions, for example, through investigating the rationality behind assumptions (see for example, LeBlanc 2012). What is accepted without being challenged, that is the unrecognised assumptions, which premise the discussion, can be considered the basis for this common sense. However, unpacking how choices are made in the media about what can and what cannot be discussed involves considering who has power and who does not. McLaren (1989) summarised this perspective by stating:

> The dominant culture tries to 'fix' the meaning of signs, symbols, and representations to provide a 'common' worldview, disguising relations of power and privilege through organs of mass media, state apparatus such as schools, government institutions, and state bureaucracies. (p. 174)

Research on mediatisation of educational policy can provide insights into how policy about mathematics is being shaped by politicians as common sense through ensuring only certain topics get discussed (Stack 2006), while others are rendered as being beyond dispute. For example, using Radford's definition of common sense, it is possible to see Sjøberg's (2015) discussion, of the acceptance by the media of PISA results as evidence of a country's economic well-being, as an example of the media constructing a new common sense. Similarly, in Sweden, media discussion, which extended school results into discussions about what should occur in early childhood institutions can also be considered as constructing a new common sense. "The role of the media is important for understanding the focus on school readiness as a result of their interest in the development of Swedish pupils' school results as measured by the OECD's international assessments (PISA)" (Jönsson et al. 2012, p. 5). By focussing on school results, what already occurs in early childhood institutions, the old common sense, can be ignored as irrelevant as what has become important is how to prepare children for school and international test taking. Thus, the media becomes complicit in politicians' need to present policy decisions as common sense.

In this chapter, we focus on how a Norwegian Minister of Education constructed his argument so that it appears as common sense. To do the analysis, we use the ideas from Edelman's (1988) political spectacle which have been used in previous research on the use of media to transform educational policy (Anderson 2005, 2007; Hattam et al. 2009; Smyth 2006). Before describing the methodology, we present the case.

2   The Case

The reason for pursuing this particular example is that the Norwegian barnehage, alongside its equivalent institutions in the other Nordic and northern European countries, traditionally has been firmly rooted in the "social policy pedagogical tradition" (Bennett 2005).

> Staff are trained to work in open framework contexts, and structural conditions support an active learning approach. The guiding national curriculum is flexible enough to allow staff to experiment with different pedagogical approaches, and adapt programmes to local conditions and demand. Again, Nordic guidelines are formulated on a consultative basis, and receive the critical analysis and consent of the major stakeholders before becoming statutory. (Bennett 2005, p. 11)

However, there are indications that this approach to early childhood education is under pressure to change to the "readiness for school tradition" (Bennett 2005), which emphasises learning specific content, such as mathematics. This approach is teacher-directed with a focus on child outputs, often assessed during the programme to ensure easy transition to school. Bennett (2005) made the distinction between these two approaches as a consequence of analysing the curriculum documents for early childhood institutions in 20 countries. From this analysis, he identified these two broad traditions, although most curricula combined features from both. Drawing on the OECD (2001) document *Starting Strong*, Norway, along with other Nordic countries, was considered to have a strong commitment to the social pedagogy tradition. However, even at that time, Bennett (2005) noted that these countries were instituting requirements for children to participate in more academically-focused activities.

> Increasingly, OECD countries regard early childhood as a period in which children should be introduced to literacy and numeracy. Economic and labour market reasons drive this focus to some extent, as literacy, numeracy and technology proficiency are fast becoming indispensable in modern economies, with many service sector jobs now requiring high standards of reading comprehension and analysis. (Bennett 2005, pp. 15–16)

In Norway, the last year of barnehage, for 6 years old, was shifted to become the first year of school in 1997 (Hansen and Simonsen 2001). This did alleviate, to some degree, anxieties about young children's transitions to school, as this first year of school was designed to ease them into school routines. It also meant that the barnehage curriculum, known as the *Framework Plan for the Content and Tasks of Kindergartens* (Kunnskapsdepartementet 2011), could retain its strong connection to the social policy pedagogy approach, as it continues to be based on "holistic pedagogical philosophy, with care, play and learning being at the core of activities" (Jensen 2009, p. 12).

Since then, the tendency to pay greater attention in barnehage to educationally significant goals has increased. This has resulted in a substantial amount of research into how mathematics learning opportunities could be incorporated into the social pedagogy tradition (see for example the work of Scandinavian mathematics education researchers documented in Meaney et al. 2016). However, the traditional consultative approach to forming barnehage policy, described by Bennett (2005), seems to have been replaced by politicians using the media to change public opinion before policy is proposed. The lack of consultation means that the voices of researchers as well as other key stakeholders are left without possibilities for providing input into the policy.

Since coming into government in 2013 and accepting the position of Minister for Education, Torbjørn Røe Isaksen has focused on strengthening the teaching of mathematics and science, known as "realfag" in Norwegian. This focus may have been connected to the Prime Minister, Erna Solberg's claim when in opposition that if she gained power, she would improve Norway's position in the testing program of the Organisation for Economic Co-operation and Development (OECD) (Sjøberg 2015). Regarding mathematics in barnehage, from October 2013, when the government was elected, till April 2016, his Ministry has released 64 press releases and the Minister has written about 200 debate articles on this topic. The debate article that we analyse here was written about 22 months after he came into government. As is the case with this one, debate articles were sometimes written after one of his policy proposals was criticised. This suggests that he was conforming to a recognised media strategy of politicians of providing quick rebuttal to any criticism (Franklin 2004).

In August 2014, the Norwegian Minister for Education gave a short interview in a regional newspaper, *Bergens Tidende* ("Norske elevar gir opp for lett" 2014), in which he first agreed with a lower secondary school teacher's description of students giving up too quickly when they perceived mathematics problems as being hard. He then went on to promote a forthcoming strategy for improving students' interest and achievement in realfag. The strategy addresses, among other issues, how mathematics teaching could start earlier. The Minister added, "perhaps we should introduce realfag already in barnehage".

A local barnehage teacher, Tone Digranes, picked up on this last remark and— on the grounds that science and mathematics constitute two out of seven knowledge areas in the framework plan for barnehage (Kunnskapsdepartementet 2011)—attacked the Minister for lack of knowledge of the barnehage curriculum (Digranes 2014).[2] In his answer two days later, the Minister defended himself by referring to Fröbel's[3] geometrical toys, his "gifts", and his experiences from visiting several barnehager, and then argued for a stronger focus on mathematics in barnehage (Isaksen 2014). He has reiterated this same argument many times since he became Minister and it has become a central platform in his Ministry's attempts at being seen to improve education in Norway. The Minister's argument was:

> I believe that an even stronger emphasis on maths [in barnehage] can be a good measure to reverse the trend of poor maths performance in school. / Mathematics is the school subject that causes the students the biggest problem. Not only are many Norwegian students at a low level in mathematics, but there are also few who score high. ... Bad results in mathematics can have serious consequences for the individual student – it is actually so that the grades one get in maths and science have the greatest impact on whether one manages to complete *videregående*[4] [upper secondary school]. At the same time, the number of doctorates in mathematics and science decreases. For society, this is serious. Norway needs scientific expertise to develop new technologies and to secure its well-being in the future. Innovation, research and the use of high technology require that we have a certain number of people with top competence in mathematics and other sciences. / Therefore, realfag is one of the government's main priorities. We need a new culture of realfag, and nothing is better than awakening interest in realfag already in barnehage. (Isaksen 2014; our translation)

Building on a model of economic progress based on scientific development, the minister's argument for a stronger focus on mathematics in barnehage is composed as a chain of six cause-and-effect claims:

- Norway's well-being in the future will come from innovation, research and use of high technology
- Innovation, research and use of high technology will come from an increase in the number of doctorates in mathematics and science

---

[2]The rubric of Digranes' reply "Kunnskapsløst av kunnskapsministeren" is a pun in Norwegian apposing "knowledge-lessness" (i.e. ignorance) with the Norwegian title for the Minister for education which translates to "Minister for knowledge".

[3]Friedrich Fröbel is known as the father of kindergartens, having set up the first ones in Germany in the nineteenth century. His "gifts" were a set of toys that support children's learning, including the learning of mathematical ideas.

[4]We have consistently provided the names of the Norwegian school system in Norwegian to indicate, like barnehage, that English terms do not provide the nuances of the Norwegian system.

- The number of doctorates in mathematics and science will increase with less dropouts in videregående (upper secondary)
- Less dropouts in videregående (upper secondary) will occur if there are fewer students with low exam grades at the end of *grunnskole* (lower secondary school) and more high-achievers
- Fewer students with low exam grades at the end of grunnskole (lower secondary school) and more high-achievers will be the result of awakened interest in mathematics in barnehage
- Awakened interest in mathematics in barnehage will occur as a result of a stronger focus on mathematics in barnehage.

The short version of the argument is that more mathematics in barnehage results in the society's future well-being, specifically the economic well-being. Suffice to say, the whole argument is not stronger than the weakest of the links in the chain. For example, Drori's (2000) extensive research showed that more emphasis on science education in curricula did not automatically lead to economic progress. If there was a relationship, it was that the more curriculum emphasis was linked to the least amount of economic progress. There is therefore no evidence to show that emphasis on mathematics curricula in barnehage will yield the economic benefits that the Minister suggested.

However, rather than focusing on the validity of the argument, we analyse it as an instance of mediatisation of education policy and consider how the Minister used particular discursive resources to present his argument as common sense. We then use this analysis to discuss how the Minister subtly shifts the focus on mathematics in barnehage to the readiness for school approach and away from the social pedagogy tradition, so that opportunities for consultation seem unnecessary.

3 Analysing Media Production of Policy

Rodney et al. (2016) described different methodologies used by researchers in Canada to investigate media discussions, including about mathematics education. These approaches include framing (Barwell and Abtahi 2015), critical discourse analysis (LeBlanc 2012) and positioning theory which Rodney et al. (2016) used. Chorney et al. (2016) who had an article in the same journal as Rodney et al. (2016) and drew on the same set of media reports also used positioning theory. Of these earlier research studies, LeBlanc's (2012) has the most similarities with our research, as it was also concerned with the mediatisation of policy discussions. However, we have chosen to use the ideas of Murray Edelman, a political scientist, to analyse what strategies the Minister used to construct his argument rather than critical discourse analysis, because Edelman's work focused specifically on how politicians utilised the media.

Murray Edelman wrote several books (e.g. 1964/1985, 1977, 1985, 1988) in which he explored the social psychology of politics and the consequences for democracy of the "political spectacle", his term for the pervasive reporting of news in readily available media. The discussions in the books are extensive and long ranging, making it difficult to identify specific points. However, Anderson (2005, 2007) extracted from Edelman's work six strategies that could be seen in the use of media by politicians to both present, but also to construct educational policy. Although Anderson (2005) noted that the media were just one contributor to political elites' construction of "political consensus around 'ruling ideas'" (Edelman 1988, p. 199) in the political spectacle, Anderson was able to exemplify their overt use by politicians in different circumstances to ensure that the general public came to take specific perspectives for granted, that is as common sense. It was this connection to common sense and the use by other researchers (see Miller-Kahn and Smith 2001; Smyth 2006) of Edelman's ideas in understanding the role of media in education policy debates that made it clear that his ideas would be valuable for our research.

Anderson (2007) listed the six strategies as:

- Importance of language and discourse
- The definition of events as crises
- A tendency to cover political interests with a discourse of rational policy
- The linguistic evocation of enemies and the displacement of targets
- The public as political spectators
- The media as mediator of the political spectacle. (pp. 108–109)

In the next sections, we introduce each of these strategies and analyse the Minister's argument to determine whether and how the Minister utilised these strategies to construct his argument for why barnehager should include more mathematics. In this short, one-quarter page debate piece, not all six strategies were equally evident. Nonetheless, we found it surprising that so many of them were present, particularly as Edelman's work had been situated in another field, political science, and in another country, USA.

3.1   Importance of Language and Discourse

From Edelman's perspective language and discourse were particularly important in setting the problem (Smyth 2006). As Anderson (2007) summarised, the importance of language and discourse is concerned with what Edelman (1977) aptly phrased as "the linguistic structuring of social problems" (p. 26) and "how the problem is named involves alternative scenarios, each with its own facts, value judgments, and emotions" (p. 29). Consequently, the choice of words and images channel the public into seeing a situation in a specific way through connecting to a particular sets of values (Anderson 2007).

In the debate article, we argue that the Minister used specific terms, about Fröbel and his gifts, to situate himself as knowledgeable about barnehage to a specific audience, barnehage teachers. This provided the opportunity to quickly rebut Digranes' (2014) critique about his lack of awareness of the barnehage framework plan. Franklin (2004) indicated that the British Labour government at the end of the 1990s deliberately provided quick rebuttals in the media to the voicing of any opposition to their policies. Isaksen's response two days after Digranes' (2014) criticism can be considered to be such a rebuttal, in that it tried to silence any dissent to his view and ministerial authority. The use of references to Fröbel indicates that the rebuttal was directed at the barnehage teacher, Diagranes, and other educators, as these references might not be understood by the general public. Although he also mentioned that he had visited barnehager and seen children engaging in mathematics and science, this is unlikely to invoke the same reassurance to barnehage teachers as the name of the founder of barnehage and his well-known geometrical toys, his gifts. As Edelman (1977) wrote, "the authoritative status of the source makes his or her definition of the issue more readily acceptable for an ambivalent public called upon to react to an ambiguous situation" (p. 25).

The careful choice of terms suggests that the Minister wanted to connect to a set of values from the traditions of barnehage, while also showing himself to be knowledgeable about barnehage. In this way, he indicated that his suggestion for more mathematics and science in barnehage was not to be viewed as something new or as moving barnehage in a new direction, away from the social pedagogy tradition. His article was designed to convince this audience that the common sense that he used was not so different to the common sense that they drew on in their work in barnehage. He was merely suggesting that for the economic well-being of Norway, there should be more attention given to something that already had a long historical association with barnehage.

3.2   The Definition of Events as Crises

Anderson's (2007) second strategy that was used by politicians was for them to define events as crises. For Edelman (1988) a crisis is not an inherent feature of a situation but rather something that has been manufactured between politicians and the media. "A crisis, like all new developments, is a creation of the language used to depict it; the appearance of a crisis is political act, not a recognition of a fact or of a rare situation" (p. 31). The choice of language ensures that the general public recognises a situation as a crisis, and that it is in their interests for politicians to solve, often by extraordinary means which the public, by default of not having any options, is likely to acquiesce to.

In the debate piece, the Minister described school students' mathematics results in terms of a crisis, a crisis for the individual students, but more so for society. Norway's economic well-being is dependent on more people having PhDs in realfag who could use and develop new technologies. Without this, it is implied, the Norwegian society would be in difficulties, particularly economically. This suggests that unless this crisis is dealt with immediately, there will be a larger crisis just waiting to happen with much wider consequences for society. Although resolving the crises maybe at the expense of shifting the social pedagogy tradition towards the readiness for school tradition, this shift is hidden in the presentation of the solution as being more of what is already occurring, although maybe at the expense of children's other activities in barnehage.

In tying this argument to the discourse of crisis, the Minister was using global ideas that have been circulating for some time. Working in Australia, Smyth (2006) identified a myth about education being in a crisis and which blamed schools, teachers and teacher educators. This was despite the fact that evaluations of the Australian education system could find no actual evidence of this crisis, resulting in it being labelling a crisis in confidence rather than a crisis in reality (Thomas 2002). Although not explicitly discussed in relationship to the education system, the Minister's article has an implicit link to the idea that the present system is in crisis and the failure of many young people to achieve in mathematics is one symptom of it.

Similarly, framing the crisis in terms of the economic well-being, which is difficult to argue against, also has both a historical and global spread. For example, Thomson et al. (2012) in an introductory article to a special issue of a journal on educational policy and school change stated:

> Governments around the world are committed to changing education. These changes are framed by national economic imperatives and driven by the need to be globally competitive in today's globalised economy. This is not change driven by an imaginary of a better and more socially just future for all, but of a more competitive economy, powered by improved human capital and better skills. (p. 1)

Thus, the state of affairs that the Norwegian Minister of Education implicitly referred to as a crisis was neither new nor specific to the situation in Norway. However, he was able to situate the crisis, of poor student results in mathematics, within the specific circumstances of Norway. Although the link between teenagers' test scores and the need for more mathematics in barnehage seems somewhat tenuous, it is presented as common sense, something that should be taken-for-granted, and not needing to be questioned.

In presenting the link as common sense, the Minister also indicates that school requirements of children are important considerations for barnehage. By not paying enough attention to them, barnehage risks the children's individual as well as society's well-being. In this way, priority in barnehage is implicitly shifted away from children's holistic development (Jensen 2009) as part of the social pedagogy tradition and towards their need to do well at school, the readiness for school approach (Bennett 2005).

3.3   A Tendency to Cover Political Interests with a Discourse of Rational Policy

The third strategy that Anderson (2007) identified in Edelman's writing was the tendency to cover political interests with a discourse of rational policy. The production of a crisis results in the logical conclusion that something must be done, thereby providing the government with an opportunity to interfere with aspects of education that may have previously been out of their control. "When the ideological agenda of government needs to be concealed, for example, in the desire of government to more closely and tightly control the work of teachers and schools, it is convenient to disguise the real intent" (Smyth 2006, p. 310). In order to persuade the public of the necessity and naturalness of this interference, then what is being promoted must be presented as rational.

It is possible to see the push for more mathematics as a way of more closely controlling the work done in barnehage, where traditional definitions of curriculum with planned lessons based on predetermined content have previously been rejected as inappropriate for young children (Bennett 2005). Situating barnehage as a kind of school, or pre-school, allows for the same type of government control as experienced by schools to be seen as natural (Schaanning 2015). To align the learning in barnehage with the more formal school curriculum, the Minister invoked discursive resources, which have been suggested for decades for why mathematics should be taught in schools. In this way, he situated barnehage and school as being the same kind of institution with the same kinds of purposes. These reasons situated the policies as being rational as they have long been accepted in the school circumstance. Niss (1996) summarised the typical reasons for why mathematics should be taught in schools:

Analyses of mathematics education from historical and contemporary perspectives show that in essence there are just a few types of fundamental reasons for mathematics education. They include the following:

- contributing to the *technological and socio-economic development* of society at large, either as such or in competition with other societies/countries;
- contributing to society's *political, ideological and cultural maintenance and development*, again either as such or in competition with other societies/countries;
- providing *individuals with prerequisites which may help them to cope with life* in the various spheres in which they live: education or occupation; private life; social life; life as a citizen. (Niss 1996, p. 13; original italics)

The future "technological and socio-economic development of society", as highlighted by Niss (1996), of Norway was the primary reason provided in the Minister's argument for why there should be more mathematics education in barnehage. The main purpose of proposing more mathematics in barnehage was to contribute to the development of the Norwegian society by facilitating an increase in labour force qualifications involving science and mathematics.

It is interesting to identify what the Minister chose not to use in his justification. Historically in Norway, emphasis is placed on holistic education (Sjøberg 2014) rather than the perceived societal needs for labour force qualifications. This is particularly the case in barnehage, where the barnehage goals as stated in the law relate to society's political, ideological and cultural maintenance and development as well as the children's individual life coping skills ("Barnehageloven" 2005), i.e. Niss' second and third reason. The Minister only superficially invoked Niss' third reason by referring to school students' need for mathematics qualifications. These qualifications are not connected to becoming democratic citizens, but rather are only to do with completing senior secondary school. Therefore, by highlighting economic rather than democratic needs, it seems that the Minister was attempting to shift perceptions of the purpose of Norwegian education, specifically in regards to barnehage, thus affecting what comes to be considered as common sense. The common sense, he promulgated, was that labour force qualifications formed the main reason for providing state education and that barnehage should have this as their main focus. Barnehage teachers' traditional understandings about the need to support children to become democratic citizens (Alvestad 2004) was no longer the common sense that could remain unquestioned.

3.4   The Linguistic Evocation of Enemies and the Displacement of Targets

The fourth strategy, used by politicians in their interactions with the media identified by Anderson (2007), is the identification of an enemy, or enemies, which act as a smoke-screen that shifts attention away from new policy initiatives. In the debate article, the Minister did not situate any one or institution as an enemy. Although it could be imagined that he could have situated the barnehage teacher, Tone Digranes, in this way, his language was guarded in how he responded to her criticism. Instead, he seemed to present himself as being in alignment with her perspective by providing his reasons for his earlier views. This lack of evoking enemies could be that unlike US politics, Norway's cultural values would not accept this way of addressing issues as appropriate, at least not in the context in question. It is, thus, interesting to see that other strategies were used to make the policy initiative, more mathematics in barnehage, become the accepted common sense.

3.5   The Public as Political Spectators

Those who control the media discourse are able to situate the public as political spectators, as they decide what should be discussed, how it should be discussed and by whom (Anderson 2007). In this way, the public is sidelined from participating in democratic discussions about how to resolve the issue, let alone from deciding whether there is an issue. The issue becomes a political spectacle to be watched, but not engaged in, by the public at large.

By situating himself as the determiner that there—in effect—is a crisis for Norwegian society, the Minister also situated himself as being the person who could provide the solution to the crisis. In this way, the public is restricted from engaging in the discussion about more realfag in barnehage, as who can argue against the need to be concerned with the future well-being of Norway? Researchers and other stakeholders in barnehage policy are also sidelined as the solution to the crisis has already been found.

Although online news resources can provide opportunities for comments from the general public (see for example Lange and Meaney 2014), these possibilities are often constrained by the structure of the news item. With this debate article, there were no options for public discussion, except by writing a new debate piece and hope the newspaper would publish it. In the article, the Minister situates the government as the ones in control who know what is good for the public—"Therefore, realfag is one of the government's main priorities. We need a new culture of realfag, and nothing is better than awakening interest in realfag already in barnehage" (Isaksen 2014; our translation). The public is situated, not as those who can influence, but instead as those needing to be influenced by media on the government's rational education policy. There is no need for them to become engaged because they are being cared for by the government and by him, as the Minister, in particular.

Even though this article was aimed at barnehage teachers, the latter are grouped with the general public as not having valuable contributions to make to the discussion. Digranes' criticisms are not dismissed out of hand, but merely adapted to show that they were in alignment with the Minister's own points. He knew and understood their situation. Barnehage teachers did not have to worry as the Minister was well informed about barnehage as well as the needs of the Norwegian economy. Consequently, the common sense that is being produced is that the public, including barnehage teachers, should remain outside of the difficult decision-making as political spectators and be confident that the government could determine what was best for society. As the discussion was not explicitly about shifting to a readiness for school tradition, but was rather situated as a need to do more of the same in barnehage, barnehage teachers did not need to participate in the discussion but could take on the role of spectators instead. Mathematics education researchers were similarly excluded from the discussion because the solution to the crisis had already been identified and thus there was nothing for them to contribute to.

3.6   The Media as Mediator of the Political Spectacle

As Anderson (2007) stated "the political spectacle is produced with media as its central conduit" (p. 109). Politicians use the media to present a particular version of a situation in carefully crafted language so that values and beliefs are brought to the fore and the public are channelled into accepting the common sense value of this viewpoint. The media's role is not one of examining or critiquing different perspectives. As Edelman (1988) claimed, "widening of the frame (in time, space, logic, and empirical links) within which an event is viewed would change its meaning but would also create an account typically categorized as research rather than as news and often as dull rather than dramatic" (p. 102).

Consequently, the genre of a short newspaper article may lead readers not to expect the Minister to use evidence to support his argument that more mathematics in barnehage will lead to economic well-being for Norway. The implicit message is that such evidence is not actually needed because it is assumed to be well-known or unequivocal, or because the claims are common sense. Little discussion by the public is needed about the value of the policy when its benefits are self-evident or a necessary response to a crisis.

The Minister has used the media consistently over the time he has been in office to present a particular version of the world. This debate article is merely one example of this strategy. Debate articles in the Norwegian press are important ways of communicating and raising disagreements. However, in this debate article, the Minister chose not to situate himself as being in disagreement with Digranes. Instead, he turned the genre of the debate article around to indicate that rather than being in disagreement (she was just ill-informed), they were on the same side, he merely wanted more of the same, that is more mathematics in barnehage.

The common sense that is promoted is that the media does provide opportunities for discussion and disagreement. However, as this debate article shows this role is an illusion, which the Minister can manipulate to his own advantage in order to present a shift towards the readiness for school tradition as nothing new and, thus, not requiring extensive discussion by others.

4   Conclusion

Earlier research on mediatisation of educational policy showed that politicians actively used the media to present new policy so that it appeared as common sense (Franklin 2004; Stack 2006). In doing so, they reduce the possibilities for public engagement in debates to discussions of technical issues, whereby ideological differences became hidden from view (Clarke 2012). Our contention is that the use of media by politicians sideline the role of researchers, as well as other stakeholders, as contributors to policy development. Unless researchers also learn how to become media managers, their research will have little impact, unless it is in alignment with the common sense being promoted by politicians. Providing reactive critiques are likely to achieve little response if they go against the established common sense understandings about mathematics education.

In the example provided in this article, it can be seen that discussions about the ideologies behind the social pedagogical approach and the readiness for school tradition become impossible in media discussions that emphasise only the economic well-being of Norway because of a lurking crisis caused by an insufficient supply of scientifically-skilled labour force. There is a displacement of the target of the discussion, so the shift in traditions is hidden from view. The discussion of what is needed for the well-being of Norway is in reality a Trojan horse bringing in changes that Norwegian barnehage teachers are not likely to regard as being in the best interest of children's holistic development (Alvestad 2004).

Our analysis, using the six strategies of political spectacle (Anderson 2005; Edelman 1988) showed how new "common sense" was used to position the public as spectators, whose role was to accept the benevolence of the government in providing the only appropriate solution to the crisis currently facing Norway. The findings from previous studies (Franklin 2004; Stack 2006) suggest that by using a set of typical politician media strategies, this may have been a deliberate strategy by the Minister to shift understandings about the role of barnehage. Using debate articles in the media enabled the Minister to set up his argument as sensible and in so doing create a new "common sense" for the general public. Situating arguments as common sense is a global approach used by governments (see for example, LeBlanc 2012), in which democracy becomes reframed as a spectator sport and other ideologies ignored as unimportant. In such a way, education can be situated as being primarily about labour-force requirements without public outcries. In this case, it may lead to barnehage replacing the current social pedagogy tradition with that of a readiness for school tradition without the need for stakeholder, including mathematics education researchers, to be involved in discussions about such a change. Democracy requires discussion and this includes hearing the voices of those with professional expertise, as well as those of parents and even the children whose experiences in barnehage will be altered. The mediatisation of policy could open up for wider discussions with the general public, but that is only likely to occur if politicians see their job as including listening to experts, not just presenting themselves as the ones who have the most knowledge and the best interests of society at heart.

References

Alvestad, M. (2004). Preschool teachers' understandings of some aspects of educational planning and practice related to the national curricula in Norway. *International Journal of Early Years Education, 12*(2), 83–97. https://doi.org/10.1080/0966976042000225499.

Anderson, G. L. (2005). Performing school reform in the age of the political spectacle. In B. K. Alexander, G. L. Anderson, & B. Gallegos (Eds.), *Performance theories in education: Power, pedagogy, and the politics of identity* (pp. 199–220). Mahwah: Lawrence Erlbaum.

Anderson, G. L. (2007). Media's impact on educational policies and practices: Political spectacle and social control. *Peabody Journal of Education, 82*(1), 103–120. Available from: http://www.jstor.org/stable/25594736.

Barnehageloven, Lov om barnehager: LOV-2005-06-17-64. (2005).

Barwell, R., & Abtahi, Y. (2015). Morality and news media representations of mathematics education. In S. Mukhopadhyay & B. Greer (Eds.), *Proceedings of the Eights International*

Mathematics Education and Society, 21st–26th June 2015, Portland Oregon (pp. 298–311). Portland: Mathematics Education and Society. Available from: http://mescommunity.info/.

Bennett, J. (2005). Curriculum issues in national policy-making. *European Early Childhood Education Research Journal, 13*(2), 5–23. https://doi.org/10.1080/13502930585209641.

Chorney, S., Ng, O.-L., & Pimm, D. (2016). A tale of two more metaphors: Storylines about mathematics education in canadian national media. *Canadian Journal of Science, Mathematics and Technology Education, 16*(4), 402–418. https://doi.org/10.1080/14926156.2016.1235746.

Clarke, M. (2012). The (absent) politics of neo-liberal education policy. *Critical Studies in Education, 53*(3), 297–310. https://doi.org/10.1080/17508487.2012.703139.

Digranes, T. (2014, August 7). Kunnskapsløst av kunnskapsministeren. At Torbjørn Røe Isaksen vet så lite om barnehagens innhold, er svært overraskende. *Bergens Tidende*. Retrieved from http://www.bt.no/meninger/debatt/Kunnskapslost-av-kunnskapsministeren-3172003.html.

Drori, G. S. (2000). Science education and economic development: Trends, relationships, and research agenda. *Studies in Science Education, 35*(1), 27–57. https://doi.org/10.1080/03057260008560154.

Edelman, M. (1964/1985). *The symbolic uses of politics: With a new Afterword*. Urbana IL: University of Illinois Press.

Edelman, M. (1977). *Political language: Words that succeed and policies that fail*. New York: Academic Press.

Edelman, M. (1985). Political language and political reality. *PS, 18*(1), 10–19. https://doi.org/10.2307/418800.

Edelman, M. (1988). *Constructing the political spectacle*. Chicago: University of Chicago Press.

Franklin, B. (2004). Education, education and indoctrination! Packaging politics and the three 'Rs'. *Journal of Education Policy, 19*(3), 255–270. https://doi.org/10.1080/0268093042000207601.

Gellert, U., Jablonka, E., & Keitel, C. (2001). Mathematical literacy and common sense in mathematics education. In B. Atweh, H. Forgasz, & B. Nebres (Eds.), *Sociocultural research on mathematics education* (pp. 57–73). Mahwah, NJ: Erlbaum.

Gravemeijer, K., & Doorman, M. (1999). Context problems in realistic mathematics education: A calculus course as an example. *Educational Studies in Mathematics, 39*(1–3), 111–129. https://doi.org/10.1023/A:1003749919816.

Hansen, A., & Simonsen, B. (2001). Mentor, master and mother: The professional development of teachers in Norway. *European Journal of Teacher Education, 24*(2), 171–182. https://doi.org/10.1080/02619760120095561.

Hattam, R., Prosser, B., & Brady, K. (2009). Revolution or backlash? The mediatisation of education policy in Australia. *Critical Studies in Education, 50*(2), 159–172. https://doi.org/10.1080/17508480902859433.

Hjarvard, S. (2008). The mediatization of society: A theory of the media as agents of social and cultural change. *Nordicom Review, 29*(2), 105–134.

Isaksen, T. R. (2014, August 9). Matte i barnehagen (Maths in kindergarten). *Bergens Tidende*. Retrieved from http://www.bt.no/meninger/debatt/Matte-i-barnehagen-3173406.html.

Jensen, B. (2009). A Nordic approach to early childhood education (ECE) and socially endangered children. *European Early Childhood Education Research Journal, 17*(1), 7–21. https://doi.org/10.1080/13502930802688980.

Jönsson, I., Sandell, A., & Tallberg-Bromann, I. (2012). Change or paradigm shift in the Swedish preschool? *Sociologia, Problemas e Prácticas, 69,* 47–61. Available from: http://spp.revues.org/815.

Keitel, C., & Kilpatrick, J. (2005). Mathematics education and common sense. In J. Kilpatrick, C. Hoyles, & O. Skovsmose (Eds.), *Meaning in mathematics education* (pp. 105–128). New York: Springer Science.

Kunnskapsdepartementet. (2011). *Framework plan for the content and tasks of kindergarten*. Oslo: Author [The Norwegian Ministry of Education and Research].

Lange, T., & Meaney, T. (2014). It's just as well kids don't vote: The positioning of children through public discourse around national testing. *Mathematics Education Research Journal, 26* (2), 377–397. https://doi.org/10.1007/s13394-013-0094-3.

LeBlanc, R. J. (2012). Representing new math: Genre chains and controversy in the Saskatchewan Media. *Alberta Journal of Educational Research, 58*(2), 286–299.

Lingard, B., & Rawolle, S. (2004). Mediatizing educational policy: The journalistic field, science policy, and cross-field effects. *Journal of Education Policy, 19*(3), 361–380. https://doi.org/10. 1080/0268093042000207665.

McLaren, P. (1989). *Life in schools: An introduction to critical pedagogy in the foundations of education*. London: Longman.

Meaney, T., Helenius, O., Johansson, M., Lange, T., & Wernberg, A. (Eds.). (2016). *Mathematics education in the early years: Results from the POEM2 conference, 2014*. New York: Springer International Publishing. https://doi.org/10.1007/978-3-319-23935-4.

Miller-Kahn, L., & Smith, M. L. (2001). School choice policies in the political spectacle. *Education Policy Analysis Archives, 9*(50), 1–41. Available from: http://epaa.asu.edu/ojs/issue/view/vol9.

Niss, M. (1996). Goals of mathematics teaching. In A. J. Bishop, K. Clement, C. Keitel, J. Kilpatrick, & C. Laborde (Eds.), *International handbook of mathematics education* (pp. 11–47). Dordrecht: Kluwer.

Norske elevar gir opp for lett: Kunnskapsminister Torbjørn Røe Isaksen (H) meiner elevane må lære å anstrenge seg. (2014, August 3). *Bergens Tidende*. Retrieved from http://www.bt.no/nyheter/lokalt/–Norske-elevar-gir-opp-for-lett-3170167.html.

OECD. (2001). *Starting strong: Early childhood education and care*. Paris: OECD Publishing. https://doi.org/10.1787/9789264192829-en.

Radford, L. (2008). Connecting theories in mathematics education: Challenges and possibilities. *ZDM Mathematics Education, 40*(2), 317–327. https://doi.org/10.1007/s11858-008-0090-3.

Rodney, S., Rouleau, A., & Sinclair, N. (2016). A tale of two metaphors: Storylines about mathematics education in Canadian national media. *Canadian Journal of Science, Mathematics and Technology Education, 16*(4), 389–401. https://doi.org/10.1080/14926156.2016.1235747.

Schaanning, E. (2015). Hvis skolematematikken ikke fantes. *Arr - idéhistorisk tidsskrift,* (4 Liv, Arr, idéhistorie. Festtidsskrift til Espen Schaanning). Available from: http://www.arrvev.no/artikkel/hvis-skolematematikken-ikke-fantes.

Sjøberg, S. (2014). PISA-syndromet: Hvordan norsk skolepolitikk blir styrt av OECD. *Nytt Norsk Tidsskrift, 31*(1), 30–43.

Sjøberg, S. (2015). PISA and global education governance—A critique of the project, its uses and implication. *Science & Technology Education, 11,* 111–127. https://doi.org/10.12973/eurasia.2015.1310a.

Smyth, J. (2006). The politics of reform of teachers' work and the consequences for schools: Some implications for teacher education. *Asia-Pacific Journal of Teacher Education, 34*(3), 301–319. https://doi.org/10.1080/13598660600927208.

Stack, M. (2006). Testing, testing, read all about it: Canadian press coverage of the PISA results. *Canada Journal of Education, 29*(1), 49–69. Available from: http://www.csse-scee.ca/CJE/Articles/Articles.htm.

Thomas, S. (2002). Contesting education policy in the public sphere: Media debates over policies for the Queensland school curriculum. *Journal of Education Policy, 17*(2), 187–198. https://doi.org/10.1080/02680930110116525.

Thomson, P., Lingard, B., & Wrigley, T. (2012). Ideas for changing educational systems, educational policy and schools. *Critical Studies in Education, 53*(1), 1–7. https://doi.org/10.1080/17508487.2011.636451.

"Now There's Everything to Stop You": Teacher Autonomy Then and Now

Gill Adams and Hilary Povey

Abstract Globalisation and neoliberal political agendas currently dominate educational policies and practices in, amongst others, many Anglophone and northern European countries including England, with discourses of the market and performance circulating widely and having become established regimes of truth. This demands sustained critique of hegemonic, taken-for-granted understandings and an exploration of how the lived experience of neoliberalism can be disrupted. In this chapter, we utilise the tools of genealogy to develop a history of the present, focussing particularly on the variation in autonomy revealed through a study of mathematics curriculum development. Juxtaposing stories from teachers involved in the Smile mathematics curriculum development project in England in the 1970s and 1980s with responses from currently serving teachers to the experience of performativity we highlight differences in teacher autonomy over time. We conclude by discussing the possibilities for teachers to mobilise such stories in their resistance to dominant, neo-liberal discourses.

Keywords Socio-historical · Neoliberalism · Autonomy

1 Introduction

Globalisation and neoliberal political agendas currently dominate educational policies and practices in, amongst others, many Anglophone and northern European countries including England. Discourses of the market and performance circulate widely and have become established regimes of truth, undermining teachers' professional and personal identities and placing their sense of independence, autonomy and moral authority under threat (Day and Smetham 2009). The need to critique and

G. Adams (✉) · H. Povey
Sheffield Hallam University, Sheffield, UK
e-mail: G.Adams@shu.ac.uk

H. Povey
e-mail: H.Povey@shu.ac.uk

© Springer International Publishing AG 2018
M. Jurdak and R. Vithal (eds.), *Sociopolitical Dimensions of Mathematics Education*, ICME-13 Monographs, https://doi.org/10.1007/978-3-319-72610-6_12

209

disrupt this agenda has been argued extensively elsewhere (see, for example, Ball et al. 2012; Berry 2012; Darragh et al. 2017; Llewellyn 2017; Montecino and Valero 2017) and we do not rehearse these arguments here. Rather, in this chapter, we begin to utilise a genealogical approach, to develop a "disordered and fragmentary" genealogy (Foucault 1976/1980, p. 85), or history, of mathematics teacher autonomy. Foucault defines genealogy as research that aims to activate 'subjugated' historical knowledge (O'Farrell 2005, p. 68). Here, we focus on a consideration of neoliberalism, developing a history of the present (Popkewitz 2013) illustrating teacher autonomy in different times. We begin by developing "systemic narratives" (Goodson 2014, p. 34) drawing on documentary evidence to identify historical periods of teacher autonomy and education policy in England before narrowing our focus to mathematics curriculum development. Juxtaposing stories from teachers involved in the *Smile* mathematics curriculum development project in the 1970s and 1980s with responses from currently serving teachers to the experience of performativity, we construct a conversation over time around teachers' time and energy; a focus on students; collaborative teacher learning through curriculum development; professional autonomy; and personal autonomy.

Although based on the situation in England, aspects of the policies and practices described herein will resonate with readers in many other countries, for neoliberal policies are in evidence around the world (see, for example, Darragh 2017; Goodson 2014). Our moral purpose in this chapter is to expose "intolerable taken-for-granted exercises of power" (Ball 2013a, p. 145), using stories from the past to show alternatives. In drawing on the past, we seek to disrupt the present and provoke a search for an alternative future.

2 Teacher Autonomy

Autonomy, a key feature of the complex and contested concept of professionalism in teaching and teacher development, has varied over time. Four historical phases in the changing nature of teacher professionalism and professional learning are particularly evident in Anglophone countries (Hargreaves 2000). The first, the pre-professional age, lasts until the 1960s. This was a time when teaching was seen as straightforward, common-sense and was learned through apprenticeship. From the 1960s to the mid-1980s, Hargreaves details the "age of the autonomous professional" (p. 158). During this phase, teachers "enjoyed unprecedented autonomy over curriculum development and decision making" (p. 158), traditional pedagogical approaches were questioned and there were an increasing number of progressive initiatives. However, despite this autonomy, Hargreaves cites research findings that support for teachers was limited and many remained isolated. This perspective is challenged by our account of mathematics curriculum development at that time (see Sect. 5.1).

From the 1980s to 2000, individual autonomy gave way to what Hargreaves describes as the age of the "collegial professional" (p. 162). The pace of reform accelerated and demands were made of teachers to teach in particular ways, to collaborate with colleagues and to develop new skills. However, collaboration was frequently limited to compliance with initiatives rather than any fundamental change or seen as an additional burden when working conditions did not facilitate such shared working. Looking back, Berry notes a shift from the turn of the century to autonomy "that is both directed and coercive" (Berry 2012, p. 399), autonomy that must be earned. Speculating on the fourth age, Hargreaves notes evidence of a post-professional era, where professionalism is "diminished or abandoned" (p. 167). A democratic alternative, the postmodern professional, requires teachers to build a social movement, one which by-passes governments and neoliberal polices (Hargreaves 2000, p. 175).

This reduction in autonomy is not restricted to England but part of a global trend and one which has a particular impact on mathematics teachers. Montecino and Valero (2017) analyse the ways that international agencies including the OECD and UNESCO have contributed to discourses which position the mathematics teacher as a policy product, key to improving the quality of mathematics education. In order to secure this improvement, teachers are required to engage in continuous professional development (CPD) and be subject to repeated testing: "'quality control' becomes a constant measurement that the teacher must face" (p. 144). In this environment, where "value replaces values" (Ball 2003, p. 217) teachers must attend to the enterprising self, in a quest for excellence. Any potential benefits of CPD are lost "when performativity reduces it to a set of compliance targets; points to be amassed" (Sugrue and Mertkan 2017, p. 16). In this way, the desirable attributes of the mathematics teacher are established and standardized internationally: "the teacher [is] controlled, produced and planned … subjected to the whims of the market, the development of policies, and the response to social demands" (Montecino and Valero 2017, p. 144). This control is pervasive, enacted through "meticulous, often minute, techniques" (Foucault 1979, p. 139) dictating not only the curriculum, the structure of a lesson, the approach to teaching a particular concept, but also the focus of an individual teacher's learning and even the clothing they must wear.

3 Towards a History of the Present

The genesis for this chapter was our engagement with a socio-historical study of the *Smile mathematics* curriculum development project, a teacher-led project which began in London in the 1970s. The study was based on the conviction that "history is about the present" (Hodgkin and Radstone 2003, p. 1). We have argued elsewhere (Povey and Adams 2017a) that the looking backward which the study involved was not backward-looking but, rather, forward-looking. Our historical interest is "present-minded" (Samuel 1980, p. 168) seeking to develop an understanding of subjective experience and everyday social relationships [that] can be used to pose major questions in politics and theory, and to transform our understanding of some of the leading phenomena of our time. (Samuel 1980, 173–174)

To gain purchase on that subjective experience and the associated everyday social relationships, we have collected vivid personal accounts using shared memory. In addition, we have collected a variety of materials that have enabled us to begin to build an archive detailing the *Smile* project, focussing on the period 1972–1990. Such documentary evidence forms a "particular, local, regional knowledge" (Foucault 1976/1980, p. 82).

In this chapter, we employ the tools of genealogy to better understand changes in teacher autonomy over time. Genealogy works with subjugated historical knowledges, both "historical contents that have been buried and disguised" (Foucault 1976/1980, p. 81) and "knowledges that have been disqualified as inadequate to their task or insufficiently elaborated" (p. 82). These latter, popular knowledges may take the form of teacher testimony of the kind collected in the *Smile* project. Popular knowledges are important, for "it is through the re-appearance of this knowledge, of these local popular knowledges, these disqualified knowledges, that criticism performs its work" (p. 82).

4   A Brief History of Education in England 1944–2017

4.1   *"Optimism and Trust"*

In order to develop an understanding of the social and political conditions in which *Smile* was first conceived and later flourished, we need to go further back in time, beyond its beginnings in the 1970s. The 1944 Education Act established secondary education for all in a "school system that reflected the values of a democratic society" (Newsam 2016, p. 180). The Act was drafted "during a war against totalitarian governments in which institutions like schools, and what was taught inside them, were directly controlled by the government" (Newsam 2016, pp. 180–181). In an effort to ensure such central control was not possible in England, the Act required agreement between the Local Education Authority and national government before any publically funded school could be opened, closed or changed in character. In a period of post-war economic growth, this was a "national system, locally administered" (Chitty 2009, p. 115). Although the Act provided free schooling for all children of secondary age, they were to be separated at age 11 by so called measures of ability and aptitude and directed to either grammar, technical, or modern schools. London was one of a few authorities choosing not to adopt a tri-partite secondary school system (in practice this was typically a bi-partite system of grammar and secondary modern schools), instead setting out a plan for the development of comprehensive schools. The bi-partite system perpetuated class divisions with around 80% of children from mainly working class backgrounds educated in the more poorly resourced secondary modern schools. By 1958 there were 26 comprehensive schools in London (Ball 2013b, p. 77), a considerable number given that less than 5% of the secondary school population in England was educated in comprehensive schools at that time (Chitty 2009, p. 29).

It is important to note that the Act provided no guidance on curriculum content (Chitty 2009). Peter Newsam, Chief Education Officer of the Inner London Education Authority (ILEA) from 1975 to 1981, sees the omission of curriculum guidance as deliberate, for "[i]t was not seen as the role of local or central government in a democratic society to require schools to teach pupils particular things in any particular way" (Newsam 2016, p. 182). This understanding, that decisions concerning curriculum and teaching were the responsibility of individual school staff, continued until the late 1970s (Chitty 2009). Brighouse dubs the period from the 1944 Education Act up to the early 1970s, a time where teachers had considerable freedom, as one of "optimism and trust" (Brighouse 2016, p. 153). Although Local Education Authorities provided advice to schools, "control of the curriculum and how it was taught was regarded as sacrosanct" (Brighouse 2016, p. 154), the responsibility of individual schools and of teachers.

In the 1960s there were early indications that the situation regarding teachers' autonomy over the curriculum was about to change, as Sir David Eccles (Conservative Minister of Education) raised the prospect of entering "the secret garden of the curriculum" (Chitty 2009, p. 147). Eccles established the Curriculum Study Group in 1962, a group that was viewed with suspicion by the teaching unions and replaced two years later by the more democratic Schools Council (Chitty 2009; Pring 2016). This latter organisation had teachers at the centre, working in partnership with universities to "undertake research and development work in all aspects of curriculum and examinations in primary and secondary schools" (Chitty 2009, p. 148). The Humanities Curriculum Project exemplifies such partnerships; supported by the Schools Council and building on the research of Lawrence Stenhouse (Elliot and Norris 2012), the project informed the practice of action research in schools. A 1975 paper published by the Schools Council affirms and justifies support for teachers' role in curriculum development:

> We believe the surest hope for the improvement of the secondary-school curriculum lies in the continuing professional growth of the teacher, which, in turn, implies that teachers take even greater responsibility for the development of schools curriculum policies. (Schools Council 1975, p. 30)

The active engagement of teachers was important for the success of mathematics curriculum development. During that period, there was a belief that for curriculum development to succeed it must be viewed as more than "merely the production of new syllabuses and texts" and must recognise the role, experience and understanding of the individual teacher and "encompass aims, content, methods and assessment procedures" (Howson et al. 1981, p. 2).

4.2 An Era of Reform

The year 1988 saw the introduction of the Education Reform Act for education in England and Wales and, to a lesser extent, the rest of the UK and it ushered in an era of constant reform. The Act is widely regarded as one of the most significant UK education acts in modern times, introduced by a government which "sought to drive neo-liberal principles into the heart of public policy" (Jones 2003, p. 107, quoted in Gillard 2011). By neoliberalism we understand a wide-ranging ideology in which the market is regarded as supreme, ensuring efficiency and quality in all sectors of public and private life. The domain of the state should be shrunk as small as possible with public services run by the private sector. Further, since, as Margaret Thatcher, an early exponent of neoliberalism, is famously quoted as saying "there's no such thing as society" (Thatcher 1987), individualism is valorized and encouraged to run rife. In order that the market be operationalized and individuals appropriately rewarded or disciplined, quantification and comparison become universal. Each of these features can be seen to be at work, profoundly shaping current educational policies and practices in England.

Although ostensibly about the curriculum, for mathematics education, the Education Reform Act brought very little change in curriculum content, thus making its intended purpose clearer: it was about "a centrally imposed and nationally validated system of grading children, schools and teachers" (Noss 1990, p. 28). The Act introduced universal testing into both primary and secondary schooling and, in 1992, the inspection service the Office for Standards in Education (Ofsted) was created to police the consequences for teachers and schools. This monitoring and the high stakes of the judgements that are then made about school students, individual teachers and whole schools, have consequences for teachers' identities, subjecting them to increased surveillance and reducing their independence (Day and Smethem 2009).

The "audit ideology" (Groundwater-Smith and Mockler 2009, p. 5), evident in the school inspection system and the accompanying league tables, is a key instrument in establishing a neoliberal regime of truth in education. No longer conceived of as a public good (Macpherson et al. 2014), education becomes a consumer product subject to market forces with teachers and schools measured and ranked to enable customer choice. This "epidemic of reform" changes who teachers are as well as what they do (Ball 2003, p. 215), eroding teachers' autonomy and challenging their individual and collective professional and personal identities (Day and Smethem 2009, p. 142). Ball (2003) suggests teachers are subject to the terrors of performativity and that there is a current struggle over the teacher's soul. Teachers (alongside all neoliberal subjects) are expected to 'perform' an entrepreneurial self, organizing and presenting themselves in response to targets, quality indicators, measures, scores and evaluations, crafting their identity against these parameters of success (Keddie 2016). Indeed, it is impossible to over-estimate the significance of this in the life of the school, as a complex of surveillance, monitoring, tracking, coordinating, reporting, targeting, motivating. (Ball et al. 2012, p. 525)

Currently, in England, pupil performance in mathematics examinations at age sixteen usually operates as the single most important item of data in judging secondary schools, with mathematics teachers therefore routinely experiencing greater pressure and coming under more scrutiny than most, if not all of their colleagues.

5 Mathematics Curriculum Development

5.1 Teachers and Curriculum Development 1960–1975

Curriculum development between 1960 and 1975 in England was supported by both private and public funds with teachers involved in much of the development work. Teachers' centres provided a meeting place at a local level, with access to resources, advice and in-service courses. During this period, teachers remained active in curriculum development with subject specialists offering advice but teachers taking "decisions concerning goals, content and methods" (Howson et al. 1981, p. 172). Two mathematics curriculum projects, the Fife Mathematics Project in Scotland and the School Mathematics Project (SMP) in England, illustrate variation in teachers' roles in curriculum initiatives of the time, further contextualising the subsequent discussion of the *Smile mathematics* project.

The first, the Fife Mathematics Project, developed in response to the introduction of comprehensive schools and mixed ability classes. Materials which aimed to encourage self-reliance in students and to provide opportunities for personal exploration of mathematical concepts were developed by Geoff Giles, then at Stirling University, and piloted in one secondary school prior to expansion to around 20 local schools. This project was supported by public funds, with teachers as important collaborators in developing the pedagogical approach rather than "creators of materials" (Howson et al. 1981, p. 45). Decisions around the use of the materials remained the province of individual teachers. A second project, the School Mathematics Project (SMP), was initially conceived as a research project based at the University of Southampton in the early 1960s. The SMP was funded by industry and charitable foundations, with the objective of introducing a new mathematics syllabus with materials for text-books, teachers' guides and examinations written by teachers (Cundy 1963). During this same period, the *Smile mathematics* project began and it is to this we turn now.

5.2   Smile Mathematics *1975–1990*

The *Smile mathematics* project (originally the Secondary Mathematics Individualised Learning Experiment—although this description was later challenged, the name *Smile* remained in common use) has its roots in London in the 1970s. Heads of mathematics departments or their delegates met at a conference at the Ladbroke Mathematics Centre, one of several such centres supporting mathematics teaching in London. One of the original group, writing in 1975, recalled this event:

> In the autumn of 1972 a week's conference was held at the Ladbroke Maths Centre for heads of mathematics departments. During that week John Stewart from Chelsea School, who had shown initiative in using a development of the Bertie Banks scheme, attracted enthusiastic attention. He felt his scheme had much to offer other schools and wanted a bigger team to work with. Several heads of department at the conference, including myself, had for a long time been anxious to run mixed ability schemes in their own schools but had been more easily daunted than John by the size of the task. We were very interested in working with him on a joint project and agreement was quickly reached by a group of schools to cooperate. (Gibbons 1975, p. 6)

A commitment to all-attainment teaching was one of the key drivers from the start of the project. Laurie Buxton, an ex-teacher and influential Mathematics Inspector in the Inner London Education Authority (ILEA), drew attention to the central role of teachers in creating and refining curriculum materials.

> *Smile* was certainly a happening and I am still not quite clear how it crept up on us. Odd bits of memory piece together for me how it came about. Firstly, Bertie Banks … his organisation sprang to life as he talked and I longed to visit his classroom.
>
> Later, stirrings at Ladbroke and then a surge of personalities as the original smilers burst upon us, bubbling and arguing, the cut and thrust …
>
> What is *Smile* now then? At the management and production end still perhaps something of a Frankenstein monster, but where it matters, in the school, a genuine salvation in some difficult situations. *Smile* has, unlike so many attempts at change, a really solid basis. It sprang from needs in the classroom, was constantly tested back there and developed, as all schemes should be by the teacher in the classroom. This is your genuine article - curriculum development as it should be. (Buxton 1975, p. 6)

Initially established and fostered under the auspices of the supportive ILEA, the *Smile mathematics* project was created and sustained by teachers. One year after the initial conference, some twenty schools were working together on *Smile*. Teachers from these schools were released from their teaching and responsibilities in school for one day a week, forming a working collective to create, refine and publish imaginative and inspiring mathematics curriculum materials for use in their own classrooms and beyond. This group embraced an investigative, problem-solving pedagogy. *Smile mathematics* saw itself as learner centred, giving students' considerable responsibility for organising and shaping their own learning and that of their learning community. The authority ascribed to students is apparent in archive materials where there is evidence that students' contributions and responses to

Smile activities are valued in various publications, as well as through their involvement in the process of testing out new materials in the classroom. *Smile* flourished in London from 1972 until the late eighties, supported both financially and philosophically by the ILEA. In 1990, the ILEA was abolished by Margaret Thatcher's administration; this and the beginnings of the neoliberal ascendancy led gradually and then increasingly rapidly to its demise. In the following section, we discuss curriculum changes after 1990.

5.3   Mathematics Curriculum Development in an Era of Reform: After 1990

As we saw above, in general, neoliberalism shrinks the size of the state. However, the goals of performativity through which the neoliberal subject is managed, also discussed above and below, require that the subject has an auditable framework against which she can be measured and against which she can measure herself. This has led, in the era of reform, to government involvement not just in a broad advisory outline for the curriculum but also in micro-specifying and micro-managing not only what is to be taught in schools but also how it is to be taught. This has been especially true for 'numeracy' and 'literacy' which for a time replaced the traditional mathematics and English.

In mathematics, the clearest example of this was the National Numeracy Strategy which primary (and later early secondary) teachers were required to follow from 1999. It claimed to be evidence based, instructing teachers on 'what works', in itself a radical reconceptualising of teaching as 'technicist' and de-personalised. However, its relationship to research was haphazard:

> sometimes recommendations are quite strongly underpinned, sometimes the evidence is ambiguous, sometimes there is little relevant literature, and sometimes the research is at odds with the recommendations. (Brown et al. 1998)

There were detailed 'unit plans' covering every aspect of the primary mathematics curriculum; an imposed major programme of 'top-down' training for teachers; and strict guidance on how every lesson was to be structured (a starter, a main activity and a plenary summing up). Each lesson was to address a single 'target' from the curriculum learning objectives and this was to appear on the board at the beginning of every lesson. All pupils should be able to recite the objective should visiting inspectors ask; and head teachers traversing the school on 'learning walks' came to police this, with teachers disciplined if the objective was not clearly visible throughout the lesson.

The contrast with the responsiveness to learners, the teacher and student authority and the teacher creativity and spontaneity of the earlier era could hardly be more extreme.

6 Methodology: Reconnecting with Our Past and Exploring the Present

6.1 Participants and Data Collection

In this chapter we draw on data collected as part of the socio-historical study referred to earlier which focused on *Smile mathematics* during the period 1972–1990. A key aim of the study was to create a public archive recording the development of this mathematics curriculum initiative. This online archive (https://smilemaths.wordpress.com/) uses images, stories, newsletters and other media together with extracts from conversations with some of those involved in *Smile mathematics*, including those present during its inception. These conversations took the form of unstructured group interviews with participants recruited through formal and informal mathematics education networks, including the *Smile* action group (SAG), and by means of a snowball sampling process, with contacts proposing others who had a role in the project. In this way a total of 24 potential participants were contacted with information about the project, with 19 participating in four distinct group conversations. Some of those unable to join the group conversations have contributed to the archive in other ways, for example with stories sent by email, photographs and other artefacts. Of the 19 participants, two were teaching in secondary schools and five were working in universities, often as mathematics teacher educators. The remaining participants maintained an interest in mathematics education into retirement. In advance of the meeting, participants were offered several questions that asked them to reflect upon: how they became involved in *Smile*; how they understood their role and responsibilities; the nature of authority and autonomy within *Smile*; and the links to other events of the time. These discussions involved between two and eight participants each, including the authors of this chapter, and lasted around three hours. The conversations were audio recorded and transcribed. The process of checking and returning transcripts to participants for validation and narrative analysis continues. Participants have also been encouraged to provide further personal commentaries and archive material.

Many of the *Smile* teacher participants in the study were young teachers in the 1970s and 1980s; several were closely involved in the beginnings of *Smile*, others had been introduced to *Smile* during their initial teacher education, often beginning their teaching careers in ILEA schools. Generally, as well as knowing us, they also knew one another though most had not met for many years.

Because it is important that the socio-historical study acts upon the contemporary world, alongside the collection of this historical material, we simultaneously began exploring these issues with four recently qualified teachers. The recently qualified teachers (RQTs) were just beginning their teaching careers. They were also known to us before this research began through their engagement with Masters level study as part of which they produced the writing which, with their consent, is reported here. They were asked to read a research article on performativity (Ball 2003) and then to write a personal account of what performativity meant to them in their daily working lives. Some two years later, two of these teachers responded to email prompts exploring their experience of *Smile*. In an earlier paper (Povey et al. 2016) we worked with reflections from just one of them, Rosie; here we draw in addition on data from James, Ruth and May (pseudonyms). In the data extracts below, the contributions of these recently qualified teachers (RQTs) are indicated by the acronym RQT to distinguish them from the contributions of the *Smile* teacher participants i.e. those who had been involved in *Smile* during the earlier period. The study was ethically approved through our University Ethics Committee.

6.2 Analytical Approach

In the earlier paper (Povey et al. 2016) we worked with reflections from Rosie, first, to offer phenomenological insights into her experience of performativity, that is, her first person accounts, and then to illustrate how she has been able to use the past, in this case *Smile* stories, to resist dominant, neo-liberal discourses and to assert an alternative identity and set of practices in her classroom. Her account foregrounded the current context within which teachers work and enabled us to glimpse the potential of our study.

In this chapter we juxtapose the *Smile* teachers' shared memory stories with the new teachers' writing on how performativity affects their everyday experiences of school life, with a view to highlighting changes in teachers' sense of autonomy over this historical period. Our analytic approach is influenced by the form of "layered stories" (Ely et al. 1997, p. 84). These stories might contain "fragments of information, splintered remembrances of many people, and ruptures of logic" that "braid together the layers of story that reveal the larger narrative" (Ely et al. 1997, p. 79). Layered stories may serve to illuminate the same event from the perspective of different individuals or over time. Here the multiple voices contribute to our genealogical work as data fragments illustrating aspects of mathematics teachers' work across time, bringing both depth and diversity to teachers' recollected experiences.

A risk facing any study working with teachers' memories is that "researchers will attach nostalgic projections of their own onto the teachers they study and falsely universalize their own preferred memories and nostalgias" (Hargreaves and Moore 2005, p. 137). As both authors are at once researchers and, as past *Smile* teachers, researched, this risk is one we work actively to reduce. It helps that the two authors experienced *Smile* at different times and in different ways, and hence have "different nostalgias" (Hargreaves and Moore 2005, p. 138). One important resource in countering this risk of nostalgic projection is the documentary archive of *Smile* publications, a resource that allows us to test out nostalgic recollections against contemporaneous accounts of events. Setting these accounts in the context of wider socio-historical evidence further helps to guard against distorting the past. Although we deliberately began our study with a focus on *Smile mathematics*, our subsequent exploration of other mathematics curriculum initiatives of the time provides us with an alternative perspective from which to gain some critical distance from *Smile*. One further way in which we work to guard against nostalgic accounts is through a commitment to sharing work in progress, our proposals for working with data, and our early writing from the project to a critical audience.

7   A Conversation About Autonomy Across Time

In this section we present fragments of stories illustrative of key themes that emerged during analysis: teachers' time and energy; a focus on students; collaborative teacher learning through curriculum development; professional autonomy; and personal autonomy. Frequently these themes were initially identified by the *Smile* teachers as they reflected on differences between their work with *Smile* materials in schools and their knowledge of mathematics teaching and learning today. The significance of these themes was confirmed through our analysis as we worked between the stories of different eras and sought additional historical accounts to deepen our understanding.

7.1   Teachers' Time and Energy

Time emerged as a significant theme in our early work with the recently qualified teachers, apparent for two teachers, Rosie and May, teaching in very different schools. At the time of our work with Rosie she was teaching in a school with a relatively privileged intake, one which was perceived (and perceives itself) as high-performing and as manifesting high 'standards'. Rosie highlighted the way in which demands of performativity absorb huge amounts of teacher time and energy.

> The sheer amount of work involved causes a significant dilemma … I have to sacrifice a huge amount of my time in order to do my job, [but] much of this is dedicated to monitoring performance and meeting targets, not improving the learning experience of my students. (Rosie, RQT)

May teaches at a school operating in more challenging circumstances. Most of the children she teaches come from backgrounds where disadvantage is experienced in one way or another. She sees the professional value of record keeping and the way in which this can provide a reflective space in which to consider the learning trajectories of individual students. However, she is also all too aware of the way that the current data-demands drain teachers' time and energy:

> Whilst writing this I have been thinking a lot about opportunity costs … I cannot help thinking that the opportunity cost of the time spent entering data into various software is time lost on planning engaging lessons … the mindless typing of one set of data into two programmes in order to send one off to be analysed by the higher ups is a waste of an hour. An hour that could have been spent on modifying a lesson. (May, RQT)

For the teachers who had been involved in the *Smile* project, time was perceived differently. Energy was invested in meaningful and productive work, in activity that the teachers valued.

> ...when we worked, when we were generating ideas and revising *Smile* cards. We were saying this is what kids do. We were anticipating what kids can do. The teachers we work with now, this is a revelation for them. Often they don't have much time for planning, but the time they have is, 'well this activity should be okay, this one should be okay and I've got a lesson and a series of activities'. Actually, they've never really thought through, they haven't got the time. Time is so precious; it's taken up with so many other things that they haven't got time to think through, to anticipate kids' responses. But that is exactly what we did when we were doing *Smile*. (fragment from group conversation, *Smile* teachers, 2016)

These *Smile* teachers talk enthusiastically of after school meetings, working weekends and conferences, noting the hard work, the challenge and the enjoyment. In contrast, the new teachers talk of time lost to 'mindless' tasks. One of the *Smile* teachers described this change. Departmental meetings had been seen as a time for "doing mathematics ... creating units ... working collaboratively as a department" under the guidance of a subject leader. Now they had gradually become taken over by "ticking boxes to fill in parts of the SEF [Self-Evaluation Form—a requirement of Ofsted]. You were supposed to talk about something that somebody else already knew the answer to" (fragments from group conversation, *Smile* teachers, 2016).

7.2 A Focus on Students

From the outset of *Smile* students were firmly at the centre. The scheme offered the flexibility for students to take responsibility for their learning, working with their teachers to select activities and extending these activities to develop understanding, reflecting their own interests. In our research conversations, the *Smile* teachers reflected on the pleasure of planning, "thinking about individual kids and how excited they might get from [a particular] card", recognising this process as "a very special thing, because you have to hold that child in your head to do it". This planning for individuals was part of a pedagogical approach centred on supporting children to engage meaningfully with mathematics, develop understanding and take "responsibility for all sorts of aspects of their learning" (extracts from group conversation, *Smile* teachers, 2016).

The recently qualified teachers had varying degrees of exposure to *Smile* through their initial teacher education and their Masters study, often exploring mathematics from the starting point of a *Smile* resource and supported by ex-*Smile* teachers, including ourselves. This provided an alternative to their own experience of learning mathematics at school and helped them to consider what learning might look like from the perspective of their own students.

> The lessons we experienced at university really inspired me ... they showed me the excitement of discovery and how that can be incorporated into teaching ... They also showed me a new approach to teaching mathematics, one that is more involved and

engaging than I had experienced as a learner before … It is something that I keep in mind now as I plan for my own classes … I know that when I look through the activities I will find activities that will suit how I want to teach my students. (Rosie, RQT)

Rosie is able to make connections with the past, the resources tangible remnants of alternative practice, refocussing attention on students' meaning making.

A lot of the tasks are investigative and allow the students to discover relationships themselves, but all of them help foster deeper understanding of why things are happening … I have a deep affection for [the *Smile* resources] because their complete focus on teaching for understanding is something that is really important to me … I can get [the students] to explore an area of mathematics themselves and discover something. (Rosie, RQT)

May, in common with many (most?) teachers retains a commitment to the centrality of her students and of her relationship with them, seeing it as the most important aspect of the job. However, this is constructed rather differently from students being at the centre of the mathematics itself.

Especially with the students that I work with, mainly from deprived backgrounds, even if the task was amazing, they would not do it without it being proposed/set up in an engaging way with a teacher that they had some measure of like and respect for. (May, RQT)

In addition, when she tries to see this commitment through, she is sometimes thwarted:

…it was decided that I would organise some form of maths trip … I wanted to take/offer it to all of the … lower sets … but this idea was rejected. I had to offer it only to [designated low-SES] students because they were identified by Ofsted last year as not making enough progress and not having enough provided for them and this trip ticks the maths intervention box. (May, RQT)

And James, one of the recently qualified teachers, finds that his students have now become understood as the bearers of targets against which he is measured and his pay is determined.

Drawing on a story from the archive (Adams and Povey 2016, pp. 85–86), we contrast this with the freedom and willingness to respond to students as individuals experienced by the *Smile* teachers.

The other thing is a lot of the theory that's being forced out now is this idea that children progress like this. If you've taught any length of time you get a kid who's sat there like this and you think for goodness' sake make some progress. It can be for ages, and then suddenly things seem to fall into place and they go shooting up. … one of the things I came across not long ago reminded me, it's called Maths Mag, and this was a boy … who said I don't like maths. I'm artistic, I'm arty. I don't like maths. He used to come back after school in Year 8 and he produced Maths Mag. This was all his work and they were little maths problems, sequencing problems and he'd do the diagrams and this, that and the other. I don't know, I suppose it would be a stencil on a Banda machine or something, would run it off and it would go out to the students …

It was maths and yes, you made sure he was still doing some sort of other work, but this is what he enjoyed doing, he wanted to publish Maths Mag. I think he did something like three versions of it, although I've only found one of them. But for him to come back after school and doing something that, as he said, "I hate maths." He didn't see that as maths. (Eades 2017)

We pick up this story again in later section.

7.3 Collaborative Teacher Learning Through Curriculum Development

Smile mathematics resources were created and revised by teachers working collaboratively, typically at writing weekends, often working in "groups of five or six preparing packages of materials" (Splash 1978). This collaboration, whether at the *Smile* Centre, at working weekends or conferences, naturally influenced the way that mathematics departments worked together in their schools:

> ...we were constantly being involved in things to the extent that we would take that as a model when we were doing our own in-house things. We would work together to create resources for **our** classrooms, rather than for **my** classroom. It became a model that we were using that gradually faded as time went by, which I thought was a shame. (extract from group conversation, *Smile* teachers, 2016)

Collaboration, both within subject departments and between schools, was a feature of mathematics curriculum development at the time, particularly in the ILEA.

> It's one of the differences between now and then that within schools subject development was much stronger then. It was a period when teachers could get out on subject development, could get involved in *Smile* and then there was a period when it seemed to me that schools closed in on themselves and development was very much about the school and not about the individual departments within the school. ...Departments became less important in terms of development and therefore teachers more and more worked as individuals rather than as a whole department. (extract from group conversation, *Smile* teachers, 2016)

The *Smile* teachers talked about curriculum development as a collective endeavor. *Smile* activities were trialled in the classroom at various stages of development and students encouraged to provide feedback. Their feedback was frequently reported in the newsletter *Splash* and contributed to the confidence that teachers had in the resources.

> That's the thing, isn't it? I think that was the great thing. When you had a *Smile* card that worked, you knew that lots of people had put a lot of energy into making sure that happened and were going to review it at some point. Things constantly were recycled and that's what I really liked, and that's what I miss enormously. (extract from group conversation, *Smile* teachers, 2016)

The problem solving approach to teaching mathematics, although new to Rosie (and, we argue, to her peers), is not new at all. Rather, it is the product of iterations of teacher-designed resources together with a broad, collaboratively developed pedagogy.

The *Smile* teachers recognise that when they discuss *Smile* they are also talking about their own development and that collaboration was a vital part of this "It was about people working together. That's what made it special, for me anyway, and inspired me and enthused me and made me think differently about teaching and learning" (extract from group conversation, *Smile* teachers 2016). It is difficult to find spaces for such collaboration today. May refers to her experience of "this every man for himself mentality" and Ruth writes that,

> as a teacher you need to be aware that judgments are being made by not only known observers but by colleagues on a daily basis.

It is very difficult to find the space for collaboration, mutual support and joint teacher learning in such a climate.

7.4   Professional Autonomy

Many of those interviewed as part of the study remain engaged in mathematics education, some working in schools or universities, some as independent advisors, others recently retired. These roles provide them with experience of mathematics teaching in England today; during our conversations they reflect on the changes, comparing their experiences as *Smile* teachers with teaching conditions today. Here we offer three fragments from these conversations.

> I was in my enthusiasm bouncing back from a *Smile* conference and having the metre cube, do you remember the framework you had that made a metre cube? I have a series of photographs... you see [the students] working at their cards and then looking up and then putting everything away, picking the tables up, stacking them, putting the chairs away, constructing them into metre cubes and then all doing the piling in, bundling into the metre cube and all the rest of it, standing around discussing it, doing it, the metre cube collapsing and kids all over the floor and then putting it all away and putting the chairs and tables back and so on and sitting back down again. To me that sums up what *Smile* was, that you could have that flexibility. ...You had that flexibility if you needed to, switching from class activity to individual activity or group activity. You had complete flexibility...(extract from group conversation, *Smile* teachers, 2016)

The photographs described above (and included in the archive https:// smilemaths.wordpress.com/in-the-classroom/cuboids/) illustrate the investigative approach that came to characterize *Smile*, providing evidence of classroom experimentation and risk-taking.

> ...you can compare what was happening in *Smile* classrooms then ... and it's quite a different comparison with now, where the amount of stuff that comes down from above, the senior management thing, every lesson must start with a hook, every lesson must start with a question, every lesson must have three cross-curricular themes and two bits of literacy... [Back then] you had autonomy within your classroom. You had expectations within the department, but I don't remember much from above that. (extract from group conversation, *Smile* teachers, 2016)

Later in the same conversation, another teacher continues this theme:

> I think a lot of the people teaching now don't remember how incredibly autonomous we were when we were teaching. ... [the standard three part lesson] became part of just the

furniture so fast, and I think an awful lot of stuff that we would think is non-autonomous is so part of the furniture that people feel autonomous. ...So I think they do have less autonomy, but they haven't noticed because it's been a bit drip, drip, drip, a bit like when you put a crab into water and heat it up gradually and eventually they boil to death. (fragment from group conversation, *Smile* teachers, 2016)

Whilst it may be true that many teachers are unaware of the reduction in teacher autonomy, sharing stories from the past can draw teachers' attention to it. As we explored how we might work with the *Smile* data, we shared at a conference the Maths Mag story told above. The author of the story had reflected: "This was something again with the flexibility. There was nothing to stop you ..." (Eades 2017). The notion that "there was nothing to stop" teachers from responding flexibly to an individual student's needs provoked one participant at the conference to respond "now there's everything to stop you" (Adams and Povey 2016, p. 86). May refers to the "*we know best* control" she experiences and Ruth, one of the recently qualified teachers, struggled in making decisions about her teaching as the knowledge that she was constantly judged by some unintelligible system left her in a state of semi-paralysis:

I have found that it is often hard to prioritize teaching tasks, never being confident as to which aspects are valued most and upon which the greatest judgements will be made, or which judgements will even be evidenced for that matter. (Ruth, RQT)

In this climate of suspicion and lack of trust teachers are unable to experiment, adopting instead the language of accountability and associated targets. Sugrue and Mertkan note how such language "gains currency through its pervasive presence; legitimacy through use" (2017, p. 15).

7.5 Personal Autonomy

It is evident from the *Smile* fragments in the sections above that, as well as professional autonomy, the *Smile* teachers experienced a high level of personal involvement, pleasure and satisfaction from their work. A strong sense of an engaged self comes across binding together the personal, the political (for which see the web archive) and the professional.

What a wonderful time we had, we really did. Didn't we enjoy ourselves ... Nobody thinks about making teachers' jobs enjoyable these days. (fragment from group conversation, *Smile* teachers, 2016)

One *Smile* teacher talked of voluntarily attending working weekends, noting the pleasure in curriculum development work.

It was really exciting to be working with other teachers trying to do something different in classrooms. (Paechter 2017)

In contrast, the new teachers talk of sacrifice and pressure, of constant comparison and judgment and of the struggle to have "a healthy life" (Rosie, RQT).

Any pleasure or satisfaction these teachers may gain from their work is difficult to discern, as the following extract illustrates.

> Teachers now are responsible for making sure they are meeting the myriad of criteria to prove to others – and themselves – that they are a good teacher. Having to constantly prove themselves drives teachers to invest huge amounts of time and energy into their job. The feeling of being constantly judged by uncertain criteria heightens the stress levels. All together it leads to a teacher who constantly questions their own ability to do their job and faces a daily personal battle over doing a good job and getting swallowed up by their work. … Teachers may have responsibility for their own performance but they have very little control over it and, if they are anything like me, feeling that you are constantly chasing a moving target and coming up short. (Rosie, RQT)

Changes in personal autonomy were also highlighted by some of the photographs included in the *Smile* archive. One striking and unexpected feature that those viewing the archive have responded to is the manner in which these teachers of the 1970s and 1980s are dressed. One is pictured wearing a t-shirt. Today, it is common in England for teachers to be expected, sometimes required, to wear 'business dress', and not unknown for them to be forbidden to cross the corridor without wearing a jacket, the individual teacher "carefully fabricated" (Foucault 1979, p. 217) in a new social order.

8 Discussion

In the socio-historical study upon which this chapter draws, we have begun to re-create a rich picture, not only of a particular curriculum development project but also of the working lives of mathematics teachers, past and present. These narratives of individuals' experiences of teaching are complemented by "systemic narratives" (Goodson 2014, p. 34), bringing together a collection of materials documenting the story of *Smile*, relating this to national developments in mathematics. Each story, fragment or extract from the archive may serve as a prompt to question existing understandings of policy and practice; by considering these alongside the reflections of recently qualified teachers we deliberately draw attention to the differences. Our moral purpose in this chapter in disrupting these taken-for-granted understandings is to unsettle. For those involved in *Smile* from the beginning, the autonomy they enjoyed was perhaps unnoticed, simply 'the way things were'. Now, as the *Smile* teachers reflect on changing conditions there is evidence of how "the space for inventiveness, experimentation and, indeed, autonomous decision making by teachers, becomes increasingly closed down" (Berry 2012, p. 400). Their stories provide all teachers with possibilities, opening up spaces for them to imagine (and begin to work towards) alternative teacher selves.

Precariousness, "a fundamental condition of the neo-liberal society" (Ball 2013a, p. 134) is evident in the comments of the new teachers in our study, the shifting values making it difficult to know to what they need to attend. Their stories highlight changes in teacher autonomy over time, standing in contrast to those of the *Smile* teachers of earlier times. Meticulous interaction with trifling data requirements act to discipline teachers (Ball et al. 2012, p. 523) through demanding their attention to minutiae, "a political anatomy of detail" (Foucault 1979, p. 139). Such Foucauldian disciplining leaves teachers with less personal resource with which to engage in a creative and moral way with the fundamental purposes of education.

The focus on 'quality' judgements diverts teachers' attention from the moral purpose of teaching. Trapped in an endless quest for progress, teachers compete with themselves and against others, leaving little time or energy to engage critically or meaningfully with each other or with their wider role. "Collective interests are replaced by competitive relations and it becomes increasingly difficult to mobilize workers around issues of general significance" (Ball 2013a, p. 135).

9 Concluding Comments

Our purpose in the socio-historical project was not merely to set past practice from the *Smile mathematics* project against present, through the use of narrative fragments and stories, but rather to offer an opportunity for the mathematics teacher today to re-appraise who she is becoming. Like Sachs (2001), we see the potential in supporting teachers to develop an activist identity through the construction of self-narratives. Such work is an important step towards "re-story[ing] themselves in and against the audit culture" (Stronach et al. 2002, p. 130). Sharing these narratives may be a first step in reclaiming social spaces where teachers might come together to reflect critically on the policy environment, in the context of this chapter an environment in which there is everything to stop independent and autonomous behaviour by teachers.

Educational research is increasingly colonised by accountability measures (Llewellyn 2017; Ball 2013a, b), now, more than ever there is a need to ensure that teachers' voices are heard. It is not our intention to present '*the* teacher's voice', an idealised, representative voice and we acknowledge that the teachers' voices in this study are "selectively appropriated ones" (Hargreaves and Moore 2005, p. 131). Nonetheless, such voices have a story to tell, one that has, until now been silent. Thus they contribute to the theme of this volume, the aim of bringing these "voices from the margin into the mainstream". As discussed above, knowledge from teachers' testimonies is hidden from history, visible at the margins if at all. By foregrounding the voices of teachers from the past we seek to (re)create spaces for teachers of today.

Working on the socio-historical study brought many surprises, resurfacing lost memories and prompting a reappraisal of past and present, a sense of nostalgia. Frequently viewed negatively, nostalgia was originally used to describe a psychological disorder, but is now considered to have multiple definitions (Sedikides et al. 2008; Zembylas 2011). A critically reflective nostalgia, one which "cherishes shattered fragments of memory and temporalizes space" (Boym 2001, p. 49 quoted in Zembylas 2011, p. 643) may, as Zembylas argues, offer opportunities for transformation. In this chapter, we see how nostalgia may also provide a reminder of what is possible, thus providing teachers with a chance to see beyond the present.

We have written elsewhere (Povey and Adams 2017b) of our hope that the socio-historical study of *Smile mathematics* will, in some small way, work to alleviate the sense of "historic loneliness" (Berger 2016a, p. 17) that is part of the neo-liberal project and that acts to disconnect us from our individual and collective pasts. There is some evidence to support this hope, both in the connections that were rekindled between the original *Smile* teachers and the optimism that the new teachers drew from the stories. The reflections on teacher autonomy presented in this chapter are intended to contribute to that wider purpose. Our intention is to work with this history to challenge the discourses of neo-liberalism, even as they work to peel us apart, increasing loneliness (Monbiot 2016). History, a meeting place and provider of company (Berger 2016b), is also a provocation to swim against "the tides of compliance, instrumentalism, fundamentalism and neo-liberalism which so categorise the contemporary age" (Groundwater-Smith and Mockler 2009, p. 139).

Acknowledgements We are grateful to the British Academy/Leverhulme for financial support for the project (Grant SG150824), to all those who have contributed so generously to the archive and to those new teachers, particularly Rosie Everley, who have given freely of their time to help us in this project.

References

Adams, G., & Povey, H. (2016). Workshop report: Using data from a history of *Smile* to overcome 'historic loneliness'. *Proceedings of the British Society for Research in Learning Mathematics, 36*, 2. Loughborough University, Loughborough, June 2016. http://www.bsrlm.org.uk/wp-content/uploads/2016/11/BSRLM-CP-36-2-15.pdf. Accessed January 10, 2017.

Ball, S., Maguire, M., Braun, A., Perryman, J., & Hoskins, K. (2012). Assessment technologies in schools: 'Deliverology' and the 'play of dominations'. *Research Papers in Education, 27*(5), 513–533.

Ball, S. J. (2003). The teacher's soul and the terrors of performativity. *Journal of Education Policy, 18*(2), 215–228.

Ball, S. J. (2013a). *Foucault, power, and education*. London: Routledge.

Ball, S. J. (2013b). *The education debate* (2nd ed.). Bristol: Policy Press.

Berger, J. (2016a). History is a meeting place. *New Statesman*, January 15–21, 2016, 17.

Berger, J. (2016b). How to resist a state of forgetfulness. In *Confabulations* (pp. 133–143). UK: Penguin.

Berry, J. (2012). Teachers' professional autonomy in England: Are neo-liberal approaches incontestable? *Forum, 54*(3), 397–410.

Brighouse, T. (2016). From 'optimism and trust' to 'markets and managerialism'. In R. Pring & M. Roberts (Eds.), *A generation of radical educational change: Stories from the field* (pp. 153–166). Abingdon, Oxon: Routledge.

Brown, M., Askew, M., Baker, D., Denvir, H., & Millett, A. (1998). Is the national numeracy strategy research-based? *British Journal of Educational Studies, 46*(4), 362–385.

Buxton, L. (1975). *How did it all happen? Splash 0010*. London: Pimlico School.

Chitty, C. (2009). *Education policy in Britain* (2nd ed.). Basingstoke: Palgrave Macmillan.

Cundy, H. M. (1963). The school mathematics project. *The Mathematical Gazette, 47*(359), 20–21.

Darragh, L. (2017). Fears and desires: Researching teachers in neoliberal contexts. In A. Chronaki (Ed.), *Proceedings of the 9th International Conference of Mathematics Education and Society: Vol.1. Mathematics Education and Life at Times of Crisis* (pp. 227–231).http://mes9.ece.uth.gr/portal/. Accessed May 26, 2017.

Darragh, L., Björklund Boistrup, L., Valero, P., Adams, G., & Povey, H. (2017). Neoliberalism: A crisis for mathematics education? In A. Chronaki (Ed.), *Proceedings of the 9th International Conference of Mathematics Education and Society: Vol. 1. Mathematics Education and Life at Times of Crisis* (pp. 149–153). http://mes9.ece.uth.gr/portal/. Accessed May 26, 2017.

Day, C., & Smethem, L. (2009). The effects of reform: Have teachers really lost their sense of professionalism? *Journal of Educational Change, 10*(2–3), 141–157.

Eades, J. (2017) Maths mag. In H. Povey & G. Adams (Eds.), *Smilemaths*. https://smilemaths.wordpress.com/themes/learning-together/students-at-the-centre/. Accessed May 12, 2017.

Elliot, J., & Norris, N. (Eds.). (2012). *Curriculum, pedagogy and educational research: The work of Lawrence Stenhouse*. London: Routledge.

Ely, M., Vinz, R., Downing, M., & Anzul, M. (1997). *On writing qualitative research: Living by words*. London: Falmer Press.

Foucault, M. (1979). *Discipline and punish: The birth of the prison*. London: Penguin.

Foucault, M. (1980). Two lectures. Lecture one: 7 January 1976. In C. Gordon (Ed.), *Power-knowledge: Selected interviews and other writings, 1972–1977* (Vintage Books ed., pp. 78–92). New York: Harvester Press.

Gibbons, R. (1975). *Smile history*. Splash 0001. London: Ladbroke Mathematics Centre.

Gillard, D. (2011). *Education in England: A brief history*. www.educationengland.org.uk/history.

Goodson, I. (2014). *Curriculum, personal narrative and the social future*. Abingdon: Routledge.

Groundwater-Smith, S., & Mockler, N. (2009). *Teacher professional learning in an age of compliance: Mind the gap*. Milton Keynes, UK: Springer.

Hargreaves, A. (2000). Four ages of professionalism and professional learning. *Teachers and Teaching, 6*(2), 151–182.

Hargreaves, A., & Moore, S. (2005). Voice, nostalgia, and teachers' experiences of change. In F. Bodone (Ed.), *What difference does research make and for whom?* (pp. 129–140). New York, NY: Peter Lang.

Hodgkin, K., & Radstone, S. (2003). Introduction: Contested pasts. In K. Hodgkin & S. Radstone (Eds.), *Contested pasts: The politics of memory* (pp. 1–22). London: Routledge.

Howson, G., Keitel, C., & Kilpatrick, J. (1981). *Curriculum development in mathematics*. Cambridge: Cambridge University Press.

Keddie, A. (2016). Children of the market: Performativity, neoliberal responsibilisation and the construction of student identities. *Oxford Review of Education, 42*(1), 108–122.

Llewellyn, A. (2017). Technologies of (re)production in mathematics education research: Performances of progress. In H. Straehler-Pohl, N. Bohlmann, & A. Pais (Eds.), *The disorder of mathematics education: Challenging the sociopolitical dimensions of research* (pp. 153–169). New York: Springer.

Macpherson, I., Robertson, S., & Walford, G. (2014). An introduction. In I. Macpherson, S. Robertson, & G. Walford (Eds.), *Education, privatisation and social justice: Case studies from Africa, South Asia and South East Asia* (pp. 9–24). Oxford: Symposium Books.

Monbiot, G. (2016, October 12). Neoliberalism is creating loneliness. That's what's wrenching society apart. *The Guardian*.

Montecino, A., & Valero, P. (2017). Mathematics teachers as products and agents: To be or not to be. That is the point! In H. Straehler-Pohl, N. Bohlmann & A. Pais (Eds.), *The disorder of mathematics education: Challenging the sociopolitical dimensions of research* (pp. 135–152). Switzerland: Springer.

Newsam, P. (2016). 1944–2015: Towards the nationalisation of education in England. In R. Pring & M. Roberts (Eds.), *A generation of radical educational change: Stories from the field* (pp. 179–190). Abingdon, Oxon: Routledge.

Noss, R. (1990). The national curriculum and mathematics: A case of divide and rule. In P. Dowling (Ed.), *Mathematics versus the national curriculum* (pp. 13–32). London: Taylor & Francis.

O'Farrell, C. (2005). *Michel Foucault*. London: Sage.

Paechter, C. (2017). Biography. In H. Povey & G. Adams (Eds.), *Smilemaths*. https://smilemaths. wordpress.com/biographies/carrie-paechter/. Accessed May 25, 2017.

Popkewitz, T. (2013). The sociology of education as the history of the present: Fabrication, difference and abjection. *Discourse: Studies in the Cultural Politics of Education, 34*(3), 439–456.

Povey, H., & Adams, G. (2017a). Thinking forward: Using stories from the recent past in mathematics education in England. In A. Chronaki (Ed.), *Proceedings of the 9th International Conference of Mathematics Education and Society: Vol. 2. Mathematics Education and Life at Times of Crisis* (pp. 803–811). http://mes9.ece.uth.gr/portal/index.php. Accessed May 26, 2017.

Povey, H., & Adams, G. (2017b). *Possibilities for mathematics education? Aphoristic fragments from the past*. Paper under review for a special issue of The Disorder of Mathematics Education.

Povey, H., Adams, G., & Everley, R. (2016). *"Its influence taints all": Mathematics teachers resisting performativity through engagement with the past*. Paper presented at 13th International Congress on Mathematical Education (ICME13), Hamburg, July 24–31, 2016.

Pring, R. (2016). Evolution of teacher training and professional development. In R. Pring & M. Roberts (Eds.), *A generation of radical educational change: Stories from the field* (pp. 81–94). Abingdon, Oxon: Routledge.

Sachs, J. (2001). Teacher professional identity: Competing discourses, competing outcomes. *Journal of Education Policy, 16*(2), 149–161.

Samuel, R. (1980). On the methods of history workshop: A reply. *History Workshop Journal, 9*(1), 162–176.

Schools Council (Great Britain). (1975). *The whole curriculum, 13–16: The report of the Schools Council working party on the whole curriculum 1971–74*. Working paper 53. London: Evans & Methuen.

Sedikides, C., Wildschut, T., Arndt, J., & Routledge, C. (2008). Nostalgia: Past, present, and future. *Current Directions in Psychological Science, 17*(5), 304–307.

Splash. (1978). *Smile* writing week. In H. Povey & G. Adams (Eds.), *Smilemaths*. https:// smilemaths.wordpress.com/themes/developing-the-curriculum/writing-weekends/. Accessed May 12, 2017.

Stronach, I., Corbin, B., McNamara, O., Stark, S., & Warne, T. (2002). Towards an uncertain politics of professionalism: Teacher and nurse identities in flux. *Journal of Education Policy, 17*, 109–138.

Sugrue, C., & Mertkan, S. (2017). Professional responsibility, accountability and performativity among teachers: The leavening influence of CPD? *Teachers and Teaching: Theory and Practice, 23*(2), 171–190.

Thatcher, M. (1987). *Interview for women's own 23 September 1987*. http://www. margaretthatcher.org/document/106689. Accessed July 26, 2017.

Zembylas, M. (2011). Reclaiming nostalgia in educational politics and practice: Counter-memory, aporetic mourning, and critical pedagogy. *Discourse: Studies in the Cultural Politics of Education, 32*(5), 641–655.

Author Index

Subject Index

© Springer International Publishing AG 2018
M. Jurdak and R. Vithal (eds.), *Sociopolitical Dimensions of Mathematics
Education*, ICME-13 Monographs, https://doi.org/10.1007/978-3-319-72610-6